LONDON MATHEMATICAL SOCIETY LECTURE NOTE SERIES

Managing Editor: Professor J.W.S. Cassels, Department of Pure Mathematics and Mathematical Statistics, University of Cambridge, 16 Mill Lane, Cambridge CB2 1SB, England

The titles below are available from booksellers, or, in case of difficulty, from Cambridge University Press.

34 Representation theory of Lie groups, M.F. ATIYAH *et al*
46 p-adic analysis: a short course on recent work, N. KOBLITZ
50 Commutator calculus and groups of homotopy classes, H.J. BAUES
59 Applicable differential geometry, M. CRAMPIN & F.A.E. PIRANI
66 Several complex variables and complex manifolds II, M.J. FIELD
69 Representation theory, I.M. GELFAND *et al*
76 Spectral theory of linear differential operators and comparison algebras, H.O. CORDES
77 Isolated singular points on complete intersections, E.J.N. LOOIJENGA
83 Homogeneous structures on Riemannian manifolds, F. TRICERRI & L. VANHECKE
86 Topological topics, I.M. JAMES (ed)
87 Surveys in set theory, A.R.D. MATHIAS (ed)
88 FPF ring theory, C. FAITH & S. PAGE
89 An F-space sampler, N.J. KALTON, N.T. PECK & J.W. ROBERTS
90 Polytopes and symmetry, S.A. ROBERTSON
92 Representation of rings over skew fields, A.H. SCHOFIELD
93 Aspects of topology, I.M. JAMES & E.H. KRONHEIMER (eds)
94 Representations of general linear groups, G.D. JAMES
95 Low-dimensional topology 1982, R.A. FENN (ed)
96 Diophantine equations over function fields, R.C. MASON
97 Varieties of constructive mathematics, D.S. BRIDGES & F. RICHMAN
98 Localization in Noetherian rings, A.V. JATEGAONKAR
99 Methods of differential geometry in algebraic topology, M. KAROUBI & C. LERUSTE
100 Stopping time techniques for analysts and probabilists, L. EGGHE
104 Elliptic structures on 3-manifolds, C.B. THOMAS
105 A local spectral theory for closed operators, I. ERDELYI & WANG SHENGWANG
107 Compactification of Siegel moduli schemes, C-L. CHAI
108 Some topics in graph theory, H.P. YAP
109 Diophantine analysis, J. LOXTON & A. VAN DER POORTEN (eds)
110 An introduction to surreal numbers, H. GONSHOR
113 Lectures on the asymptotic theory of ideals, D. REES
114 Lectures on Bochner-Riesz means, K.M. DAVIS & Y-C. CHANG
115 An introduction to independence for analysts, H.G. DALES & W.H. WOODIN
116 Representations of algebras, P.J. WEBB (ed)
118 Skew linear groups, M. SHIRVANI & B. WEHRFRITZ
119 Triangulated categories in the representation theory of finite-dimensional algebras, D. HAPPEL
121 Proceedings of *Groups - St Andrews 1985*, E. ROBERTSON & C. CAMPBELL (eds)
122 Non-classical continuum mechanics, R.J. KNOPS & A.A. LACEY (eds)
125 Commutator theory for congruence modular varieties, R. FREESE & R. MCKENZIE
126 Van der Corput's method of exponential sums, S.W. GRAHAM & G. KOLESNIK
128 Descriptive set theory and the structure of sets of uniqueness, A.S. KECHRIS & A. LOUVEAU
129 The subgroup structure of the finite classical groups, P.B. KLEIDMAN & M.W. LIEBECK
130 Model theory and modules, M. PREST
131 Algebraic, extremal & metric combinatorics, M-M. DEZA, P. FRANKL & I.G. ROSENBERG (eds)
132 Whitehead groups of finite groups, ROBERT OLIVER
133 Linear algebraic monoids, MOHAN S. PUTCHA
134 Number theory and dynamical systems, M. DODSON & J. VICKERS (eds)
135 Operator algebras and applications, 1, D. EVANS & M. TAKESAKI (eds)
136 Operator algebras and applications, 2, D. EVANS & M. TAKESAKI (eds)
137 Analysis at Urbana, I, E. BERKSON, T. PECK, & J. UHL (eds)
138 Analysis at Urbana, II, E. BERKSON, T. PECK, & J. UHL (eds)
139 Advances in homotopy theory, S. SALAMON, B. STEER & W. SUTHERLAND (eds)
140 Geometric aspects of Banach spaces, E.M. PEINADOR & A. RODES (eds)
141 Surveys in combinatorics 1989, J. SIEMONS (ed)
144 Introduction to uniform spaces, I.M. JAMES
145 Homological questions in local algebra, JAN R. STROOKER
146 Cohen-Macaulay modules over Cohen-Macaulay rings, Y. YOSHINO
147 Continuous and discrete modules, S.H. MOHAMED & B.J. MÜLLER
148 Helices and vector bundles, A.N. RUDAKOV *et al*
149 Solitons, nonlinear evolution equations and inverse scattering, M. ABLOWITZ & P. CLARKSON
150 Geometry of low-dimensional manifolds 1, S. DONALDSON & C.B. THOMAS (eds)
151 Geometry of low-dimensional manifolds 2, S. DONALDSON & C.B. THOMAS (eds)
152 Oligomorphic permutation groups, P. CAMERON

153 L-functions and arithmetic, J. COATES & M.J. TAYLOR (eds)
154 Number theory and cryptography, J. LOXTON (ed
155 Classification theories of polarized varieties, TAKAO FUJITA
156 Twistors in mathematics and physics, T.N. BAILEY & R.J. BASTON (eds)
157 Analytic pro-*p* groups, J.D. DIXON, M.P.F. DU SAUTOY, A. MANN & D. SEGAL
158 Geometry of Banach spaces, P.F.X. MÜLLER & W. SCHACHERMAYER (eds)
159 Groups St Andrews 1989 volume 1, C.M. CAMPBELL & E.F. ROBERTSON (eds)
160 Groups St Andrews 1989 volume 2, C.M. CAMPBELL & E.F. ROBERTSON (eds)
161 Lectures on block theory, BURKHARD KÜLSHAMMER
162 Harmonic analysis and representation theory, A. FIGA-TALAMANCA & C. NEBBIA
163 Topics in varieties of group representations, S.M. VOVSI
164 Quasi-symmetric designs, M.S. SHRIKANDE & S.S. SANE
165 Groups, combinatorics & geometry, M.W. LIEBECK & J. SAXL (eds)
166 Surveys in combinatorics, 1991, A.D. KEEDWELL (ed)
167 Stochastic analysis, M.T. BARLOW & N.H. BINGHAM (eds)
168 Representations of algebras, H. TACHIKAWA & S. BRENNER (eds)
169 Boolean function complexity, M.S. PATERSON (ed)
170 Manifolds with singularities and the Adams-Novikov spectral sequence, B. BOTVINNIK
171 Squares, A.R. RAJWADE
172 Algebraic varieties, GEORGE R. KEMPF
173 Discrete groups and geometry, W.J. HARVEY & C. MACLACHLAN (eds)
174 Lectures on mechanics, J.E. MARSDEN
175 Adams memorial symposium on algebraic topology 1, N. RAY & G. WALKER (eds)
176 Adams memorial symposium on algebraic topology 2, N. RAY & G. WALKER (eds)
177 Applications of categories in computer science, M. FOURMAN, P. JOHNSTONE, & A. PITTS (eds)
178 Lower K- and L-theory, A. RANICKI
179 Complex projective geometry, G. ELLINGSRUD *et al*
180 Lectures on ergodic theory and Pesin theory on compact manifolds, M. POLLICOTT
181 Geometric group theory I, G.A. NIBLO & M.A. ROLLER (eds)
182 Geometric group theory II, G.A. NIBLO & M.A. ROLLER (eds)
183 Shintani zeta functions, A. YUKIE
184 Arithmetical functions, W. SCHWARZ & J. SPILKER
185 Representations of solvable groups, O. MANZ & T.R. WOLF
186 Complexity: knots, colourings and counting, D.J.A. WELSH
187 Surveys in combinatorics, 1993, K. WALKER (ed)
188 Local analysis for the odd order theorem, H. BENDER & G. GLAUBERMAN
189 Locally presentable and accessible categories, J. ADAMEK & J. ROSICKY
190 Polynomial invariants of finite groups, D.J. BENSON
191 Finite geometry and combinatorics, F. DE CLERCK *et al*
192 Symplectic geometry, D. SALAMON (ed)
193 Computer algebra and differential equations, E. TOURNIER (ed)
194 Independent random variables and rearrangement invariant spaces, M. BRAVERMAN
195 Arithmetic of blowup algebras, WOLMER VASCONCELOS
196 Microlocal analysis for differential operators, A. GRIGIS & J. SJÖSTRAND
197 Two-dimensional homotopy and combinatorial group theory, C. HOG-ANGELONI, W. METZLER & A.J. SIERADSKI (eds)
198 The algebraic characterization of geometric 4-manifolds, J.A. HILLMAN
199 Invariant potential theory in the unit ball of $\mathbf{C}^n$, MANFRED STOLL
200 The Grothendieck theory of dessins d'enfant, L. SCHNEPS (ed)
201 Singularities, JEAN-PAUL BRASSELET (ed)
202 The technique of pseudodifferential operators, H.O. CORDES
203 Hochschild cohomology of von Neumann algebras, A. SINCLAIR & R. SMITH
204 Combinatorial and geometric group theory, A.J. DUNCAN, N.D. GILBERT & J. HOWIE (eds)
205 Ergodic theory and its connections with harmonic analysis, K. PETERSEN & I. SALAMA (eds)
206 An introduction to noncommutative differential geometry and its physical applications, J. MADORE
207 Groups of Lie type and their geometries, W.M. KANTOR & L. DI MARTINO (eds)
208 Vector bundles in algebraic geometry, N.J. HITCHIN, P. NEWSTEAD & W.M. OXBURY (eds)
209 Arithmetic of diagonal hypersurfaces over finite fields, F.Q. GOUVÊA & N. YUI
210 Hilbert C*-modules, E.C. LANCE
211 Groups 93 Galway / St Andrews I, C.M. CAMPBELL *et al*
212 Groups 93 Galway / St Andrews II, C.M. CAMPBELL *et al*
214 Generalised Euler-Jacobi inversion formula and asymptotics beyond all orders, V. KOWALENKO, N.E. FRANKEL, M.L. GLASSER & T. TAUCHER
215 Number theory, S. DAVID (ed)
216 Stochastic partial differential equations, A. ETHERIDGE (ed)
217 Quadratic forms with applications to algebraic geometry and topology, A. PFISTER
218 Surveys in combinatorics, 1995, PETER ROWLINSON (ed)
220 Algebraic set theory, A. JOYAL & I. MOERDIJK
221 Harmonic approximation, S.J. GARDINER
222 Advances in linear logic, J.-Y. GIRARD, Y. LAFONT & L. REGNIER (eds)
223 Analytic semigroups and semilinear initial boundary value problems, KAZUAKI TAIRA

London Mathematical Society Lecture Note Series. 223

Analytic Semigroups and Semilinear Initial Boundary Value Problems

Kazuaki Taira
Hiroshima University

CAMBRIDGE
UNIVERSITY PRESS

Published by the Press Syndicate of the University of Cambridge
The Pitt Building, Trumpington Street, Cambridge CB2 1RP
40 West 20th Street, New York, NY 10011-4211, USA
10 Stamford Road, Oakleigh, Melbourne 3166, Australia

First published 1995

Printed in Great Britain at the University Press, Cambridge

Library of Congress cataloging in publication data available

British Library cataloguing in publication data available

ISBN 0 521 55603 1 paperback

TABLE OF CONTENTS

Typeset by $\mathcal{A}\mathcal{M}\mathcal{S}$-TEX

TO MY MOTHER

PREFACE

This monograph is devoted to the functional analytic approach to initial boundary value problems for semilinear parabolic differential equations. First we study non-homogeneous boundary value problems for second-order elliptic differential operators, in the framework of Sobolev spaces of L^p style, which include as particular cases the Dirichlet and Neumann problems. We prove that these boundary value problems provide an example of analytic semigroups in the L^p topology. The essential point in the proof is to define a function space which is a tool well suited to investigating our boundary conditions. By virtue of the theory of analytic semigroups, one can apply this result to the study of the initial boundary value problems for semilinear parabolic differential equations in the framework of L^p spaces.

This monograph grew out of a set of lecture notes "On initial boundary value problems for semilinear parabolic differential equations" for graduate courses given at the University of Tsukuba in winter 1994/95. In order to make this monograph accessible to a broad readership, I have tried to start from scratch. In the preparatory chapters, we even prove fundamental results like a generation theorem for analytic semigroups in functional analysis and Sobolev imbeddings theorems in partial differential equations. Furthermore, we summarize the basic definitions and results about the L^p theory of pseudo-differential operators which is considered as a modern theory of potentials. The L^p theory of pseudo-differential operators forms a most convenient tool in the study of elliptic boundary value problems in the framework of Sobolev spaces of L^p style. The material in these preparatory chapters is given for completeness, to minimize the necessity of consulting too many outside references. This makes the monograph fairly self-contained.

This work was begun at the University of Turin and the University of Bologna in May 1988 under the sponsorship of the Italian "Consiglio Nazionale delle Ricerche" and a major part of the work was done at the University of the Philippines in the course of the JSPS-DOST exchange program from January 1989 to March 1989 while I was on leave from the University of Tsukuba. I take this opportunity to express my gratitude to all these institutions for their hospitality and support. Thanks are also due to the editorial staff of Cambridge University Press for their unfailing helpfulness and cooperation during the production of the book.

I hope that this monograph will lead to a better insight into the study

Typeset by $\mathcal{A}\mathcal{M}\mathcal{S}$-TEX

of initial boundary value problems for semilinear parabolic differential equations. For probabilistic information on the topics discussed here, I would like to call attention to my previous book *Boundary Value Problems and Markov Processes*, Lecture Notes in Mathematics, No. 1499, Springer-Verlag, 1991.

Higashi-Hiroshima, June 1995

Kazuaki Taira

INTRODUCTION AND RESULTS

Let Ω be a bounded domain of Euclidean space $\mathbf{R}^n$, with C^∞ boundary Γ; its closure $\overline{\Omega} = \Omega \cup \Gamma$ is an n-dimensional, compact C^∞ manifold with boundary. We let

$$A = \sum_{i=1}^{n} \frac{\partial}{\partial x_i} \left(\sum_{j=1}^{n} a^{ij}(x) \frac{\partial}{\partial x_j} \right) + \sum_{i=1}^{n} b^i(x) \frac{\partial}{\partial x_i} + c(x)$$

be a second-order *elliptic* differential operator with real C^∞ coefficients on $\overline{\Omega}$ such that:

(1) $a^{ij}(x) = a^{ji}(x)$, $x \in \Omega$, $1 \leq i, j \leq n$.
(2) There exists a constant $a_0 > 0$ such that

$$\sum_{i,j=1}^{n} a^{ij}(x)\xi_i\xi_j \geq a_0 |\xi|^2, \quad x \in \overline{\Omega},\ \xi \in \mathbf{R}^n.$$

(3) $c(x) \leq 0$ in Ω, and c does not vanish *identically* in Ω.

We consider the following boundary value problem: Given functions f and φ defined in Ω and on Γ, respectively, find a function u in Ω such that

$$(*) \qquad \begin{cases} Au = f & \text{in } \Omega, \\ Bu := a\frac{\partial u}{\partial \nu} + bu\big|_\Gamma = \varphi & \text{on } \Gamma. \end{cases}$$

Here:

(1) a and b are real-valued, C^∞ functions on Γ.
(2) $\partial/\partial\boldsymbol{\nu}$ is the conormal derivative associated with the operator A:

$$\frac{\partial}{\partial \boldsymbol{\nu}} = \sum_{i,j=1}^{N} a^{ij} n_j \frac{\partial}{\partial x_i},$$

$\mathbf{n} = (n_1, n_2, \cdots, n_n)$ being the unit exterior normal to the boundary Γ (see

Typeset by $\mathcal{A}_{\mathcal{M}}\mathcal{S}$-TEX

Figure 1).

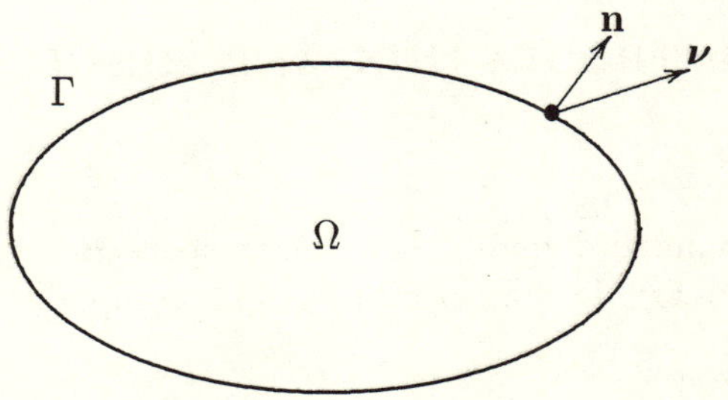

Figure 1

We remark that if $a \equiv 1$ and $b \equiv 0$ on Γ (resp. $a \equiv 0$ and $b \equiv 1$ on Γ), then the boundary condition B is the so-called Neumann (resp. Dirichlet) condition. It is easy to see that problem $(*)$ is non-degenerate (or coercive) if and only if either $a \neq 0$ on Γ or $a \equiv 0$ and $b \neq 0$ on Γ.

The first purpose of this book is to prove an existence and uniqueness theorem for problem $(*)$ under the condition $a \geq 0$ on Γ in the framework of Sobolev spaces of L^p style. The essential point is to define a function space which is a suitable tool to investigate the degenerate boundary condition B.

If $1 \leq p < \infty$, we let

$L^p(\Omega) =$ the space of (equivalence classes of) Lebesgue measurable functions u on Ω such that $|u|^p$ is integrable on Ω.

The space $L^p(\Omega)$ is a Banach space with the norm

$$\|u\|_p = \left(\int_\Omega |u(x)|^p dx \right)^{1/p}.$$

If s is a positive integer, we define the Sobolev space

$H^{s,p}(\Omega) =$ the space of (equivalence classes of) functions $u \in L^p(\Omega)$ whose derivatives $D^\alpha u$, $|\alpha| \leq s$, in the sense of distributions are in $L^p(\Omega)$.

The space $H^{s,p}(\Omega)$ is a Banach space with the norm

$$\|u\|_{s,p} = \left(\sum_{|\alpha| \leq s} \int_\Omega |D^\alpha u(x)|^p dx \right)^{1/p}.$$

Furthermore we let

$$B^{s-1/p,p}(\Gamma) = \text{the space of the boundary values } \varphi \text{ of functions } u \in H^{s,p}(\Omega).$$

In the space $B^{s-1/p,p}(\Gamma)$, we introduce a norm

$$|\varphi|_{s-1/p,p} = \inf \|u\|_{s,p},$$

where the infimum is taken over all functions $u \in H^{s,p}(\Omega)$ which equal φ on the boundary. The space $B^{s-1/p,p}(\Gamma)$ is a Banach space with respect to this norm $|\cdot|_{s-1/p,p}$; more precisely, it is a Besov space (cf. [BL], [Tr]).

We introduce a subspace of $B^{s-1-1/p,p}(\Gamma)$ which is associated with the boundary condition

$$Bu = a\frac{\partial u}{\partial \boldsymbol{\nu}} + bu\Big|_{\Gamma} = 0 \quad \text{on } \Gamma.$$

We let

$$B_*^{s-1-1/p,p}(\Gamma) = \left\{\varphi = a\varphi_1 + b\varphi_2 : \varphi_1 \in B^{s-1-1/p,p}(\Gamma),\ \varphi_2 \in B^{s-1/p,p}(\Gamma)\right\},$$

and define

$$|\varphi|^*_{s-1-1/p,p} = \inf\left\{|\varphi_1|_{s-1-1/p,p} + |\varphi_2|_{s-1/p,p} : \varphi = a\varphi_1 + b\varphi_2\right\}.$$

Then it is easy to verify that the space $B_*^{s-1-1/p,p}(\Gamma)$ is a Banach space with respect to the norm $|\cdot|^*_{s-1-1/p,p}$. We remark that the space $B_*^{s-1-1/p,p}(\Gamma)$ is an "interpolation space" between the spaces $B^{s-1/p,p}(\Gamma)$ and $B^{s-1-1/p,p}(\Gamma)$. More precisely, we have

$$B_*^{s-1-1/p,p}(\Gamma) = B^{s-1/p,p}(\Gamma) \quad \text{if } a \equiv 0 \text{ on } \Gamma,$$
$$B_*^{s-1-1/p,p}(\Gamma) = B^{s-1-1/p,p}(\Gamma) \quad \text{if } a > 0 \text{ on } \Gamma,$$

and, for general a, the continuous injections

$$B^{s-1/p,p}(\Gamma) \subset B_*^{s-1-1/p,p}(\Gamma) \subset B^{s-1-1/p,p}(\Gamma).$$

Now we can state our existence and uniqueness theorem for problem $(*)$ (cf. [Um, Theorem 1]):

Theorem 1. *Let $1 < p < \infty$ and $s > 1+1/p$. Assume that the following two conditions (H.1) and (H.2) are satisfied:*

(H.1) $a(x') \geq 0$ and $b(x') \geq 0$ on Γ.

(H.2) $b(x') > 0$ on $\Gamma_0 = \{x' \in \Gamma : a(x') = 0\}$.

Then the mapping

$$(A, B) : H^{s,p}(\Omega) \longrightarrow H^{s-2,p}(\Omega) \oplus B_*^{s-1-1/p,p}(\Gamma)$$

is an algebraic and topological isomorphism. *In particular, for any $f \in H^{s-2,p}(\Omega)$ and any $\varphi \in B_*^{s-1-1/p,p}(\Gamma)$, there exists a unique solution $u \in H^{s,p}(\Omega)$ of problem $(*)$.*

The second purpose of this book is to study problem $(*)$ from the point of view of analytic semigroup theory in functional analysis. The generation theorem for analytic semigroups is well established in the non-degenerate case in the L^p topology (cf. [Fr]). We shall generalize this generation theorem for analytic semigroups to the *degenerate* case.

First we state a generation theorem for analytic semigroups in the L^p topology. We associate with problem $(*)$ an unbounded linear operator $\mathfrak{A}$ from $L^p(\Omega)$ into itself as follows:

(a) The domain of definition $D(\mathfrak{A})$ of $\mathfrak{A}$ is the set

$$D(\mathfrak{A}) = \left\{u \in H^{2,p}(\Omega) : Bu = 0\right\}.$$

(b) $\mathfrak{A}u = Au$, $u \in D(\mathfrak{A})$.

The next theorem is an L^p version of [Ta1, Theorem 1]:

Theorem 2. *Let $1 < p < \infty$. If conditions (H.1) and (H.2) are satisfied, then we have the following:*

(i) For every $0 < \varepsilon < \pi/2$, there exists a constant $r(\varepsilon) > 0$ such that the resolvent set of $\mathfrak{A}$ contains the set

$$\Sigma(\varepsilon) = \left\{\lambda = r^2 e^{i\theta} : r \geq r(\varepsilon), -\pi + \varepsilon \leq \theta \leq \pi - \varepsilon\right\},$$

and that the resolvent $(\mathfrak{A} - \lambda I)^{-1}$ satisfies the estimate

$$\left\|(\mathfrak{A} - \lambda I)^{-1}\right\| \leq \frac{c(\varepsilon)}{|\lambda|}, \quad \lambda \in \Sigma(\varepsilon),$$

where $c(\varepsilon) > 0$ is a constant depending on ε.

(ii) The operator $\mathfrak{A}$ generates a semigroup $U(z)$ on the space $L^p(\Omega)$ which is analytic in the sector $\Delta_\varepsilon = \{z = t + is : z \neq 0, |\arg z| < \pi/2 - \varepsilon\}$ for any $0 < \varepsilon < \pi/2$ (see Figure 2).

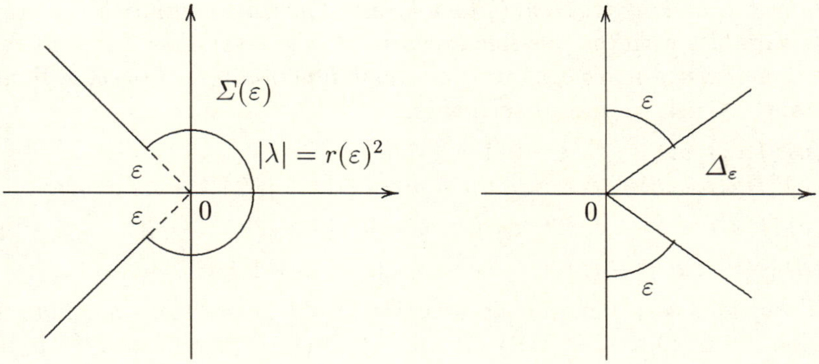

Figure 2

Next, as an application of Theorem 2, we consider the following *semilinear* initial boundary value problem: Given functions f and u_0 defined in $\Omega \times [0,T) \times \mathbf{R} \times \mathbf{R}^N$ and in Ω, respectively, find a function u in $\Omega \times [0,T)$ such that

(∗∗)

$$\begin{cases} (\frac{\partial}{\partial t} - A)u(x,t) = f(x,t,u,\operatorname{grad} u) & \text{in } \Omega \times (0,T), \\ Bu(x',t) := a(x')\frac{\partial u}{\partial \nu}(x',t) + b(x')u(x',t)\big|_{\Gamma \times [0,T)} = 0 & \text{on } \Gamma \times [0,T), \\ u(x,0) = u_0(x) & \text{in } \Omega. \end{cases}$$

By using the operator $\mathfrak{A}$, one can formulate problem (∗∗) in terms of the *abstract Cauchy problem* in the space $L^p(\Omega)$ as follows:

$$(**') \qquad \begin{cases} \frac{du}{dt} = \mathfrak{A}u(t) + F(t,u(t)), \ 0 < t < T, \\ u|_{t=0} = u_0. \end{cases}$$

Here $u(t) = u(\cdot,t)$ and $F(t,u(t)) = f(\cdot,t,u(t),\operatorname{grad} u(t))$ are functions defined on the interval $[0,T)$, taking values in the space $L^p(\Omega)$.

First we consider the case $p > n$:

Theorem 3. *Assume that conditions (H.1) and (H.2) are satisfied. Let $n < p < \infty$ and let $f(x,t,u,\xi)$ be a locally Lipschitz continuous function of all its variables with the possible exception of the x variable. Then, for every function u_0 of $D(\mathfrak{A})$, problem (∗∗′) has a unique local solution $u \in C([0,T']; L^p(\Omega)) \cap C^1((0,T'); L^p(\Omega))$ where $T' = T'(p,u_0) > 0$.*

Here $C([0,T']; L^p(\Omega))$ denotes the space of continuous functions on $[0,T']$ taking values in $L^p(\Omega)$, and $C^1((0,T'); L^p(\Omega))$ denotes the space of continuously differentiable functions on $(0,T')$ taking values in $L^p(\Omega)$, respectively.

In the case $p < n$, the domain $D(\mathfrak{A})$ is large compared with the case $n < p < \infty$. Hence we must impose some growth conditions on the function f:

Theorem 4. *Assume that conditions (H.1) and (H.2) are satisfied. Let $n/2 < p < n$ and let $f(x,t,u,\xi)$ be a locally Lipschitz continuous function of all its variables with the possible exception of the x variable. Further assume that there exist a non-negative continuous function $\rho(t,r)$ on $\mathbf{R} \times \mathbf{R}$ and a constant $1 \leq \gamma < n/(n-p)$ such that:*

(a) $|f(x,t,u,\xi)| \leq \rho(t,|u|)(1+|\xi|^\gamma)$.

(b) $|f(x,t,u,\xi) - f(x,s,u,\xi)| \leq \rho(t,|u|)\,(1+|\xi|^\gamma)\,|t-s|$.

(c) $|f(x,t,u,\xi) - f(x,t,u,\eta)| \leq \rho(t,|u|)\left(1+|\xi|^{\gamma-1}+|\eta|^{\gamma-1}\right)|\xi-\eta|$.

(d) $|f(x,t,u,\xi) - f(x,t,v,\xi)| \leq \rho(t,|u|+|v|)\,(1+|\xi|^\gamma)\,|u-v|$.

Then, for every function u_0 of $D(\mathfrak{A})$, problem $(**')$ has a unique local solution $u \in C([0,T'];\, L^p(\Omega)) \cap C^1((0,T');\, L^p(\Omega))$ where $T' = T'(p,u_0) > 0$.

Theorems 3 and 4 are a generalization of Pazy [Pa, Section 8.4, Theorems 4.4 and 4.5] to the degenerate case.

The rest of this book is organized as follows.

Chapter 1 is devoted to a review of standard topics from the theory of analytic semigroups which forms a functional analytic background for the proof of Theorems 2, 3 and 4. First we prove a generation theorem of analytic semigroups (Theorem 1.2). Next we consider the abstract linear and semilinear Cauchy problems, and give a complete proof of local existence and uniqueness theorems (Theorems 1.16 and 1.18).

In Chapter 2, we study the imbedding characteristics of Sobolev spaces of L^p style that render these spaces so useful in the study of partial differential equations. We give a complete proof of the most important of the imbedding properties of the spaces $H^{m,p}(\mathbf{R}^n)$ (Theorems 2.15 and 2.18 and Corollary 2.22).

In Chapter 3, we present a brief description of the basic concepts and results of the L^p theory of pseudo-differential operators which may be considered as a generalization of the classical potential theory. In particular, we state a Besov-space boundedness theorem (Theorem 3.17) due to Bourdaud [Bo] and a criterion for hypoellipticity for pseudo-differential operators (Theorem 3.19) due to Hörmander [Ho].

In Chapter 4, we study the boundary value problem $(*)$ in the framework of Sobolev spaces of L^p style, by using the L^p theory of pseudo-differential operators. We prove that problem $(*)$ can be reduced to the study of a pseudo-differential operator T on the boundary Γ (Theorems 4.10-4.12).

In Chapter 5, we study the pseudo-differential operator T in question, and prove Theorem 1. In Section 5.1 we prove a regularity theorem for problem $(*)$ (Theorem 5.1). More precisely, we prove that if conditions (H.1) and (H.2) are satisfied, one can construct a parametrix S for T in the Hörmander class $L^0_{1,1/2}(\Gamma)$ (Lemma 5.2), and then apply a Besov-space boundedness theorem to the parametrix S to obtain the regularity theorem for problem $(*)$. Section 5.2 is devoted to a uniqueness theorem for problem $(*)$ (Theorem 5.5). In the proof we make good use of the maximum principle stated in the

Appendix (Theorems A.1 and A.2). Section 5.3 is devoted to an existence theorem for problem (∗) (Theorem 5.7) which is an essential step in the proof of Theorem 1. To prove this, we make use of a method essentially due to Agmon [Ag] (Proposition 5.10). This is a technique of treating a spectral parameter as a second-order differential operator of an extra variable and relating the old problem to a new one with the additional variable.

Chapter 6 is devoted to the proof of Theorem 2 (Theorems 6.6 and 6.8). Once again the method of Agmon plays an important role in the proof of the surjectivity of $\mathfrak{A} - \lambda I$ (Proposition 6.7). In Section 6.1 we study the operator $\mathfrak{A}$, and prove fundamental *a priori* estimates for $\mathfrak{A} - \lambda I$ (Theorem 6.3) which play an important role in the proof of Theorem 2 in Section 6.2.

In the final chapter, Chapter 7, we prove Theorems 3 and 4. In Section 7.1 we study the imbedding properties of the domains of the fractional powers $(-\mathfrak{A})^{\alpha}$, $0 < \alpha < 1$, into Sobolev spaces of L^p style (Theorem 7.1). This allows us to solve by successive approximations the semilinear initial boundary value problem (∗∗) in Section 7.2, proving Theorems 3 and 4.

CHAPTER I

THEORY OF ANALYTIC SEMIGROUPS

This chapter is devoted to the theory of analytic semigroups which forms a functional analytic background for the proof of Theorems 2, 3 and 4. First we prove a generation theorem for analytic semigroups (Theorem 1.2). Next we consider the abstract linear and semilinear Cauchy problems, and give a complete proof of local existence and uniqueness theorems (Theorems 1.16 and 1.18). For more leisurely treatments of analytic semigroups, the reader is referred to Friedman [Fr], Pazy [Pa] and also Yosida [Yo].

1.1 Generation Theorem for Analytic Semigroups

Let E be a Banach space over the real number field $\mathbf{R}$ or the complex number field $\mathbf{C}$. Let $A : E \longrightarrow E$ be a *densely defined*, closed linear operator, that is, the domain of definition $D(A)$ of A is dense in the space E.

Assume that the operator A satisfies the following two conditions:

(1) The resolvent set of A contains the region (see Figure 1.1)

$$\Sigma_\omega = \{\lambda \in \mathbf{C} : \lambda \neq 0, |\arg \lambda| < \pi/2 + \omega\}, \quad 0 < \omega < \pi/2.$$

(2) For each $\varepsilon > 0$, there exists a constant $M_\varepsilon > 0$ such that the resolvent $R(\lambda) = (A - \lambda I)^{-1}$ satisfies the estimate

$$\text{(1.1)} \quad \|R(\lambda)\| \leq \frac{M_\varepsilon}{|\lambda|}, \quad \lambda \in \Sigma_\omega^\varepsilon = \{\lambda \in \mathbf{C} : \lambda \neq 0, \, |\arg \lambda| \leq \pi/2 + \omega - \varepsilon\}.$$

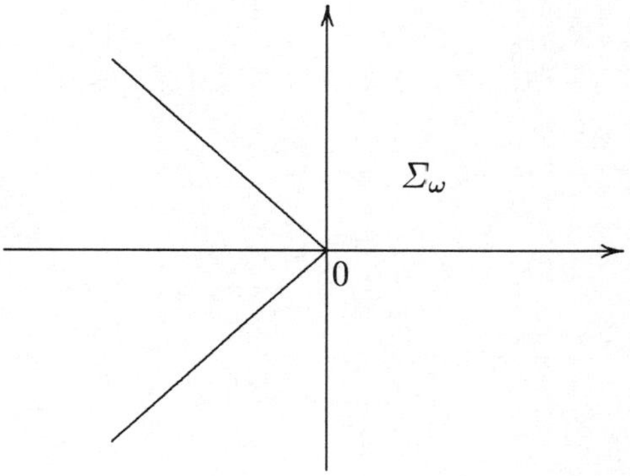

Figure 1.1

Typeset by $\mathcal{A}_\mathcal{M}\mathcal{S}$-TEX

We let

$$U(t) = -\frac{1}{2\pi i} \int_{\Gamma} e^{\lambda t} R(\lambda)\, d\lambda, \quad t > 0, \tag{1.2}$$

where Γ is a path in the set $\Sigma_{\omega}^{\varepsilon}$ consisting of the following three curves (see Figure 1.2):

$$\Gamma^{(1)} = \left\{ r e^{-i(\pi/2+\omega-\varepsilon)} : 1 \leq r < \infty \right\},$$
$$\Gamma^{(2)} = \left\{ e^{i\theta} : -(\pi/2+\omega-\varepsilon) \leq \theta \leq \pi/2+\omega-\varepsilon \right\},$$
$$\Gamma^{(3)} = \left\{ r e^{i(\pi/2+\omega-\varepsilon)} : 1 \leq r < \infty \right\}.$$

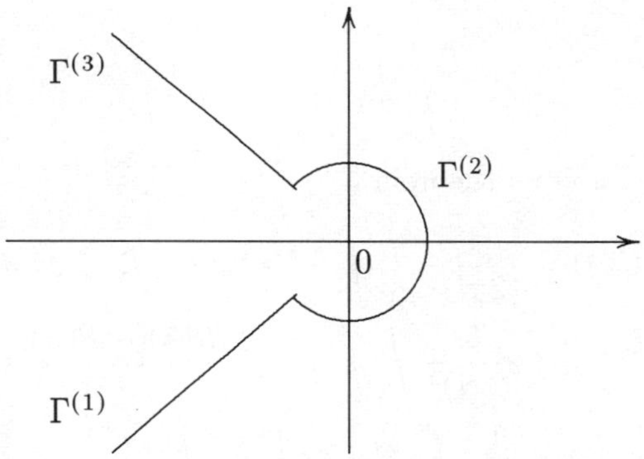

Figure 1.2

Then, by estimate (1.1), it is easy to see that the integral

$$U(t) = -\frac{1}{2\pi i} \sum_{k=1}^{3} \int_{\Gamma^{(k)}} e^{\lambda t} R(\lambda)\, d\lambda$$

converges in the Banach space $L(E, E)$ of all bounded linear operators on E into itself, for every $t > 0$, and thus it defines a bounded linear operator on E.

Furthermore, the operators $U(t)$ form a semigroup on E; more precisely, we have the following:

Proposition 1.1. *The operators $U(t)$, defined by formula (1.2), enjoy the* semigroup property, *that is,*

$$U(t+s) = U(t) \cdot U(s), \quad t, s > 0.$$

Proof. By Cauchy's theorem, one may assume that

$$U(s) = -\frac{1}{2\pi i} \int_{\Gamma'} e^{\mu s} R(\mu)\, d\mu, \quad s > 0.$$

Here Γ' is a path obtained from the path Γ by translating each point of Γ to the right by a fixed small positive distance (see Figure 1.3).

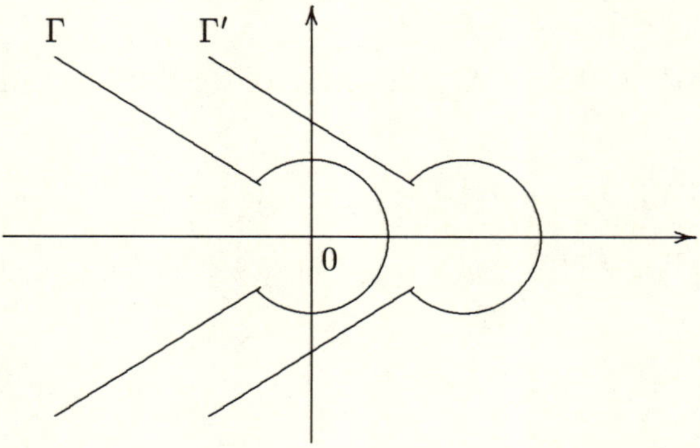

Figure 1.3

Then we have by Fubini's theorem

$$\begin{aligned} U(t)\cdot U(s) &= \frac{1}{(2\pi i)^2}\int_\Gamma \int_{\Gamma'} e^{\lambda t} e^{\mu s} R(\lambda)\, R(\mu)\, d\lambda\, d\mu \\ &= \frac{1}{(2\pi i)^2}\int_\Gamma \int_{\Gamma'} e^{\lambda t} e^{\mu s} \frac{R(\lambda)-R(\mu)}{\lambda-\mu}\, d\lambda\, d\mu \\ &= \frac{1}{2\pi i}\int_\Gamma e^{\lambda t} R(\lambda)\left[\frac{1}{2\pi i}\int_{\Gamma'} \frac{e^{\mu s}}{\lambda-\mu}\, d\mu\right] d\lambda \\ &\quad - \frac{1}{2\pi i}\int_{\Gamma'} e^{\mu s} R(\mu)\left[\frac{1}{2\pi i}\int_{\Gamma} \frac{e^{\lambda t}}{\lambda-\mu}\, d\lambda\right] d\mu. \end{aligned}$$

We calculate the two terms in the last part.

(a) We let

$$f(\mu) = \frac{e^{\mu s}}{\lambda-\mu}.$$

Then, applying the residue theorem, we obtain that (see Figure 1.4)

$$\begin{aligned} &\int_{\Gamma'^{(1)}\cap\{|\mu|\le r\}} f(\mu)\, d\mu + \int_{\Gamma'^{(2)}} f(\mu)\, d\mu + \int_{\Gamma'^{(3)}\cap\{|\mu|\le r\}} f(\mu)\, d\mu \\ &\quad + \int_{-(\pi/2+\omega-\varepsilon)}^{\pi/2+\omega-\varepsilon} f(re^{i\theta}) rie^{i\theta}\, d\theta \\ &= 2\pi i \operatorname{Res}\left[f(\mu)\right]_{\mu=\lambda} \\ &= -2\pi i e^{\lambda s}. \end{aligned}$$

But we have, as $r \to \infty$,

$$\int_{\Gamma'^{(1)}\cap\{|\mu|\le r\}} f(\mu)\, d\mu \longrightarrow \int_{\Gamma'^{(1)}} f(\mu)\, d\mu,$$

$$\int_{\Gamma'^{(3)}\cap\{|\mu|\le r\}} f(\mu)\,d\mu \longrightarrow \int_{\Gamma'^{(3)}} f(\mu)\,d\mu,$$

and

$$\left|\int_{-(\pi/2+\omega-\varepsilon)}^{\pi/2+\omega-\varepsilon} f(re^{i\theta})rie^{i\theta}\,d\theta\right| \le e^{-rs\cdot\sin(\omega-\varepsilon)} \int_{-(\pi/2+\omega-\varepsilon)}^{\pi/2+\omega-\varepsilon} \frac{d\theta}{\left|\frac{\lambda}{r}-e^{i\theta}\right|} \longrightarrow 0.$$

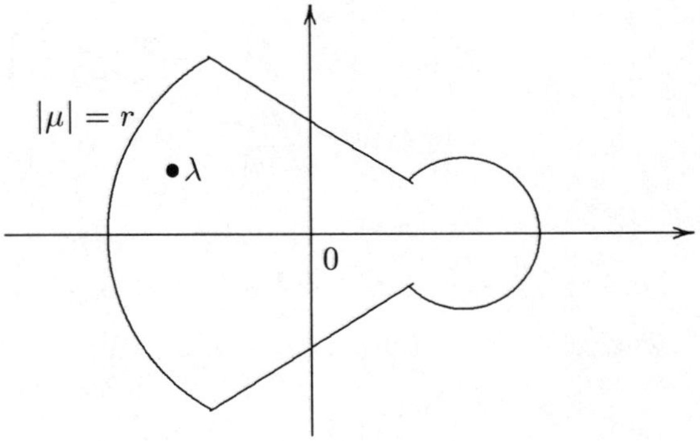

Figure 1.4

Therefore, we find that

$$\frac{1}{2\pi i}\int_{\Gamma'} \frac{e^{\mu s}}{\lambda-\mu}\,d\mu = -e^{\lambda s}.$$

(b) Similarly, since the path Γ lies to the left of the path Γ', we find that

$$\frac{1}{2\pi i}\int_{\Gamma} \frac{e^{\lambda t}}{\lambda-\mu}\,d\lambda = 0.$$

Summing up, we obtain that

$$U(t)\cdot U(s) = -\frac{1}{2\pi i}\int_{\Gamma} e^{\lambda(t+s)}R(\lambda)\,d\lambda = U(t+s).$$

The proof of Proposition 1.1 is complete. □

The next theorem states that the semigroup $U(t)$ can be extended to an analytic semigroup in some sector containing the positive real axis.

Theorem 1.2. *The semigroup $U(t)$ can be extended to a semigroup $U(z)$ which is analytic in the sector $\Delta_\omega = \{z = t + is : z \neq 0, |\arg z| < \omega\}$, and enjoys the following properties:*

(a) The operators $AU(z)$ and $\frac{dU}{dz}(z)$ are bounded operators on E for each $z \in \Delta_\omega$, and satisfy the relation

$$\frac{dU}{dz}(z) = AU(z), \quad z \in \Delta_\omega. \tag{1.3}$$

(b) For each $0 < \varepsilon < \omega/2$, there exist constants $\widetilde{M_0}(\varepsilon) > 0$ and $\widetilde{M_1}(\varepsilon) > 0$ such that

$$\|U(z)\| \leq \widetilde{M_0}(\varepsilon), \quad z \in \Delta_\omega^{2\varepsilon}, \tag{1.4}$$

$$\|AU(z)\| \leq \frac{\widetilde{M_1}(\varepsilon)}{|z|}, \quad z \in \Delta_\omega^{2\varepsilon}, \tag{1.5}$$

where

$$\Delta_\omega^{2\varepsilon} = \{z \in \mathbf{C} : z \neq 0, |\arg z| \leq \omega - 2\varepsilon\}.$$

(c) For each $x \in E$, we have as $z \to 0$, $z \in \Delta_\omega^{2\varepsilon}$,

$$U(z)x \longrightarrow x \quad \text{in } E.$$

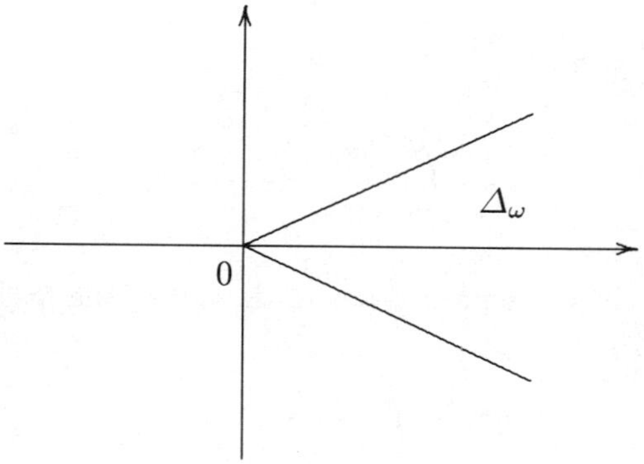

Figure 1.5

Proof. (i) The analyticity of $U(z)$: If $\lambda \in \Gamma^{(3)}$ and $z \in \Delta_\omega^{2\varepsilon}$, that is, if

$$\lambda = |\lambda| e^{i\theta}, \quad \theta = \pi/2 + \omega - \varepsilon,$$
$$z = |z| e^{i\varphi}, \quad |\varphi| \leq \omega - 2\varepsilon,$$

then we have

$$\lambda z = |\lambda|\, |z| e^{i(\theta + \varphi)},$$

with

$$\pi/2 + \varepsilon \leq \theta + \varphi \leq \pi/2 + 2\omega - 3\varepsilon < 3\pi/2 - 3\varepsilon.$$

Note that

$$\cos(\theta + \varphi) \leq \cos(\pi/2 + \varepsilon) = -\sin\varepsilon.$$

Hence it follows that

$$|e^{\lambda z}| \leq e^{-|\lambda|\,|z|\sin\varepsilon}, \quad \lambda \in \Gamma^{(3)},\ z \in \Delta_\omega^{2\varepsilon}. \tag{1.6}$$

Similarly, we have

$$|e^{\lambda z}| \leq e^{-|\lambda|\,|z|\sin\varepsilon}, \quad \lambda \in \Gamma^{(1)},\ z \in \Delta_\omega^{2\varepsilon}. \tag{1.7}$$

For each small $\varepsilon > 0$, we let

$$K_\omega^\varepsilon = \Delta_\omega^{2\varepsilon} \bigcap \{z \in \mathbf{C} : |z| \geq \varepsilon\} = \{z \in \mathbf{C} : |z| \geq \varepsilon, |\arg z| \leq \omega - 2\varepsilon\}.$$

Then, combining estimates (1.1), (1.6) and (1.7), we obtain that

$$\left\|e^{\lambda z} R(\lambda)\right\| \leq \frac{M_\varepsilon}{|\lambda|} e^{-\varepsilon \sin\varepsilon \cdot |\lambda|}, \quad \lambda \in \Gamma^{(1)} \bigcup \Gamma^{(3)},\ z \in K_\omega^\varepsilon. \tag{1.8}$$

On the other hand, we have

$$\left\|e^{\lambda z} R(\lambda)\right\| \leq M_\varepsilon e^{|z|}, \quad \lambda \in \Gamma^{(2)},\ z \in K_\omega^\varepsilon. \tag{1.9}$$

Therefore, we find that the integral

$$U(z) = -\frac{1}{2\pi i} \int_\Gamma e^{\lambda z} R(\lambda)\, d\lambda = -\frac{1}{2\pi i} \sum_{k=1}^{3} \int_{\Gamma^{(k)}} e^{\lambda z} R(\lambda)\, d\lambda \tag{1.2$'$}$$

converges in the Banach space $L(E, E)$, uniformly in $z \in K_\omega^\varepsilon$, for every $\varepsilon > 0$. This proves that the operator $U(z)$ is analytic in the domain $\Delta_\omega = \bigcup_{\varepsilon > 0} K_\omega^\varepsilon$.

By the analyticity of $U(z)$, it follows that the operators $U(z)$ also enjoy the semigroup property

$$U(z + w) = U(z) \cdot U(w), \quad z, w \in \Delta_\omega.$$

(ii) We prove that the operators $U(z)$ enjoy properties (a), (b) and (c).

(b) First, using Cauchy's theorem, we obtain that

$$U(z) = -\frac{1}{2\pi i} \int_\Gamma e^{\lambda z} R(\lambda)\, d\lambda = -\frac{1}{2\pi i} \int_{\Gamma_{|z|}} e^{\lambda z} R(\lambda)\, d\lambda,$$

where $\Gamma_{|z|}$ is a path consisting of the following three curves (see Figure 1.6):

$$\Gamma_{|z|}^{(1)} = \left\{ re^{-i(\pi/2+\omega-\varepsilon)} : \frac{1}{|z|} \leq r < \infty \right\},$$
$$\Gamma_{|z|}^{(2)} = \left\{ \frac{1}{|z|} e^{i\theta} : -(\pi/2+\omega-\varepsilon) \leq \theta \leq \pi/2+\omega-\varepsilon \right\},$$
$$\Gamma_{|z|}^{(3)} = \left\{ re^{i(\pi/2+\omega-\varepsilon)} : \frac{1}{|z|} \leq r < \infty \right\}.$$

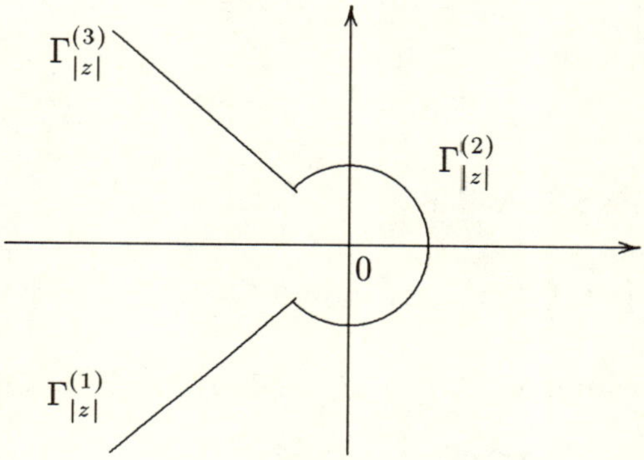

Figure 1.6

But, by estimates (1.1), (1.6) and (1.7), it follows that

$$\|e^{\lambda z} R(\lambda)\| \leq \frac{M_\varepsilon}{|\lambda|} e^{-|\lambda||z|\sin\varepsilon}, \quad \lambda \in \Gamma_{|z|}^{(1)} \bigcup \Gamma_{|z|}^{(3)}, \ z \in \Delta_\omega^{2\varepsilon}.$$

Hence we have for $k = 1, 3$

$$\begin{aligned} \int_{\Gamma_{|z|}^{(k)}} \|e^{\lambda z} R(\lambda)\| \, |d\lambda| &\leq M_\varepsilon \int_{\frac{1}{|z|}}^\infty e^{-\rho|z|\sin\varepsilon} \rho^{-1} \, d\rho \\ &= M_\varepsilon \int_1^\infty e^{-\sin\varepsilon \cdot s} s^{-1} \, ds. \end{aligned}$$

We have also for $k = 2$

$$\begin{aligned} \int_{\Gamma_{|z|}^{(2)}} \|e^{\lambda z} R(\lambda)\| \, |d\lambda| &\leq M_\varepsilon \int_{-(\pi/2+\omega-\varepsilon)}^{\pi/2+\omega-\varepsilon} e \, d\theta \\ &= 2eM_\varepsilon(\pi/2+\omega-\varepsilon) \\ &\leq 2\pi e M_\varepsilon. \end{aligned}$$

Summing up, we obtain the following estimate:

$$\|U(z)\| \leq \frac{1}{2\pi} \sum_{k=1}^3 \int_{\Gamma_{|z|}^{(k)}} \|e^{\lambda z} R(\lambda)\| \, |d\lambda|$$

$$\leq \frac{1}{2\pi}\left(2M_\varepsilon \int_1^\infty s^{-1} e^{-\sin\varepsilon\cdot s}\, ds + 2\pi e M_\varepsilon\right)$$
$$= \frac{M_\varepsilon}{\pi}\left(\int_1^\infty s^{-1} e^{-\sin\varepsilon\cdot s}\, ds + \pi e\right).$$

This proves estimate (1.4), with

$$\widetilde{M_0}(\varepsilon) = \frac{M_\varepsilon}{\pi}\left(\int_1^\infty s^{-1} e^{-\sin\varepsilon\cdot s}\, ds + \pi e\right).$$

To prove estimate (1.5), note that

$$AR(\lambda) = (A - \lambda I + \lambda I)R(\lambda) = I + \lambda R(\lambda),$$

so that

$$\|AR(\lambda)\| \leq 1 + M_\varepsilon, \quad \lambda \in \Sigma_\omega^\varepsilon.$$

Hence, arguing as in the proof of estimate (1.4), we obtain that

$$\left\|\int_\Gamma e^{\lambda z} AR(\lambda)\, d\lambda\right\| \leq 2\int_{\frac{1}{|z|}}^\infty e^{-\rho|z|\sin\varepsilon}(1 + M_\varepsilon)\, d\rho + \int_{-(\pi/2+\omega-\varepsilon)}^{\pi/2+\omega-\varepsilon} (1 + M_\varepsilon)\, e \frac{1}{|z|}\, d\theta \leq 2(1 + M_\varepsilon)\left(\int_1^\infty e^{-\sin\varepsilon\cdot s}\, ds + \pi e\right)\frac{1}{|z|}. \tag{1.10}$$

This proves that the integral $\int_\Gamma e^{\lambda z} AR(\lambda)\, d\lambda$ is convergent for every $z \in \Delta_\omega^{2\varepsilon}$. By the closedness of A, it follows that, for all $z \in \Delta_\omega^{2\varepsilon}$,

$$U(z) \in D(A),$$

and

$$AU(z) = -\frac{1}{2\pi i}\int_\Gamma e^{\lambda z} AR(\lambda)\, d\lambda. \tag{1.11}$$

Therefore, estimate (1.5) follows from estimate (1.10), with

$$\widetilde{M_1}(\varepsilon) = \frac{1 + M_\varepsilon}{\pi}\left(\int_1^\infty e^{-\sin\varepsilon\cdot s}\, ds + \pi e\right).$$

We remark that formula (1.11) remains valid for all $z \in \Delta_\omega$, since $\Delta_\omega = \bigcup_{\varepsilon>0} \Delta_\omega^{2\varepsilon}$.

(a) By estimates (1.8) and (1.9), one can differentiate formula (1.2′) under the integral sign to obtain that

$$\frac{dU}{dz}(z) = -\frac{1}{2\pi i}\int_\Gamma e^{\lambda z}\lambda R(\lambda)\,d\lambda, \quad z \in \Delta_\omega. \tag{1.12}$$

On the other hand, it follows from formula (1.11) that

$$\begin{aligned} AU(z) &= -\frac{1}{2\pi i}\int_\Gamma e^{\lambda z} AR(\lambda)\,d\lambda \\ &= -\frac{1}{2\pi i}\int_\Gamma e^{\lambda z}(I + \lambda R(\lambda))d\lambda \\ &= -\frac{1}{2\pi i}\int_\Gamma e^{\lambda z}\lambda R(\lambda)\,d\lambda, \ z \in \Delta_\omega, \end{aligned} \tag{1.13}$$

since we have by Cauchy's theorem

$$\int_\Gamma e^{\lambda z}\,d\lambda = 0.$$

Therefore, formula (1.3) follows immediately from formulas (1.12) and (1.13).

(c) Now let x_0 be an arbitrary element of $D(A)$. By the residue theorem, it follows that

$$x_0 = \frac{1}{2\pi i}\int_\Gamma \frac{e^{\lambda z}}{\lambda} x_0\,d\lambda.$$

Hence we have

$$\begin{aligned} U(z)x_0 - x_0 &= -\frac{1}{2\pi i}\int_\Gamma e^{\lambda z}\left(R(\lambda) + \frac{1}{\lambda}\right) x_0\,d\lambda \\ &= -\frac{1}{2\pi i}\int_\Gamma \frac{e^{\lambda z}}{\lambda} R(\lambda)Ax_0\,d\lambda. \end{aligned}$$

Here we remark that

$$\begin{aligned} &\left\|\frac{1}{\lambda}R(\lambda)\right\| \leq \frac{M_\varepsilon}{|\lambda|^2}, \quad \lambda \in \Gamma, \\ &\left|e^{\lambda z}\right| \leq 2e^{-|\lambda|\,|z|\sin\varepsilon} + e^{|z|}, \quad z \in \Delta_\omega^{2\varepsilon}, \ \lambda \in \Gamma. \end{aligned}$$

Thus it follows from an application of the Lebesgue dominated convergence theorem that, as $z \to 0$, $z \in \Delta_\omega^{2\varepsilon}$,

$$U(z)x_0 - x_0 \longrightarrow -\frac{1}{2\pi i}\int_\Gamma \frac{1}{\lambda}R(\lambda)Ax_0\,d\lambda.$$

But we have

$$\int_\Gamma \frac{1}{\lambda} R(\lambda) A x_0 \, d\lambda = 0.$$

Indeed, by Cauchy's theorem, it follows that

$$\begin{aligned}\int_\Gamma \frac{1}{\lambda} R(\lambda) A x_0 \, d\lambda &= \lim_{r\to\infty} \int_{\Gamma\cap\{|\lambda|\le r\}} \frac{1}{\lambda} R(\lambda) A x_0 \, d\lambda \\ &= -\lim_{r\to\infty} \int_{C_r} \frac{1}{\lambda} R(\lambda) A x_0 \, d\lambda \\ &= 0,\end{aligned}$$

where C_r is a closed path shown in Figure 1.7.

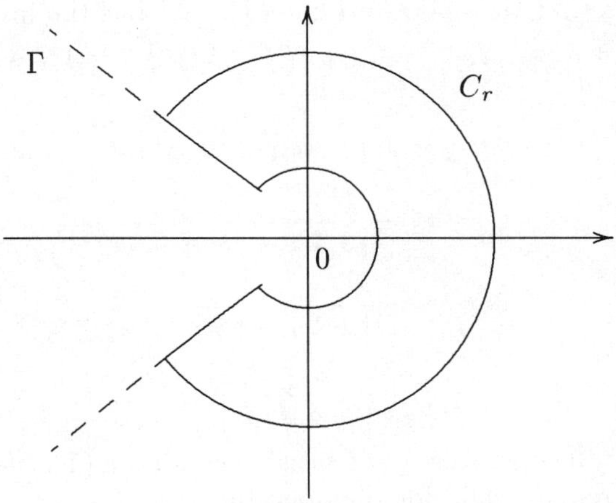

Figure 1.7

Summing up, we have proved that

$$U(z)x_0 \longrightarrow x_0 \quad \text{as } z \to 0,\ z \in \Delta_\omega^{2\varepsilon},$$

for each $x_0 \in D(A)$.

Since the domain $D(A)$ is dense in E and $\|U(z)\| \le \widetilde{M}_0(\varepsilon)$ for all $z \in \Delta_\omega^{2\varepsilon}$, it follows that, for each $x \in E$,

$$U(z)x \longrightarrow x \quad \text{as } z \to 0,\ z \in \Delta_\omega^{2\varepsilon}.$$

The proof of Theorem 1.2 is now complete. □

Remark 1.3. Assume that the operator A satisfies a stronger condition than condition (1.1)

$$\text{(1.1}'\text{)} \qquad \|R(\lambda)\| \le \frac{M_\varepsilon}{|\lambda|+1}, \quad \lambda \in \Sigma_\omega^\varepsilon.$$

Then we have the estimates

$$\|U(z)\| \leq \widetilde{M_0}(\varepsilon) e^{-\delta \cdot \mathrm{Re}\, z}, \quad z \in \Delta_\omega^{2\varepsilon}, \tag{1.4'}$$

$$\|AU(z)\| \leq \frac{\widetilde{M_1}(\varepsilon)}{|z|} e^{-\delta \cdot \mathrm{Re}\, z}, \quad z \in \Delta_\omega^{2\varepsilon}, \tag{1.5'}$$

with some constant $\delta > 0$.

Proof. Take a real number δ such that

$$0 < \delta < \frac{1}{M_\varepsilon}.$$

Then we have by estimate (1.1′)

$$\delta \left\|(A - \lambda I)^{-1}\right\| \leq \frac{\delta M_\varepsilon}{|\lambda| + 1} \leq \delta M_\varepsilon < 1, \quad \lambda \in \Sigma_\omega^\varepsilon.$$

Hence it follows that the operator $(A + \delta I) - \lambda I$ has the inverse

$$((A + \delta I) - \lambda I)^{-1} = (I + \delta(A - \lambda I)^{-1})^{-1}(A - \lambda I)^{-1},$$

and

$$\begin{aligned}
\|((A + \delta I) - \lambda I)^{-1}\| &\leq \|(I + \delta(A - \lambda I)^{-1})^{-1}\| \cdot \|(A - \lambda I)^{-1}\| \\
&\leq \frac{M_\varepsilon}{|\lambda| + 1} \frac{1}{1 - \|\delta(A - \lambda I)^{-1}\|} \\
&\leq \frac{M_\varepsilon}{|\lambda| + 1} \frac{1}{1 - \delta M_\varepsilon} \\
&\leq \left(\frac{M_\varepsilon}{1 - \delta M_\varepsilon}\right) \frac{1}{|\lambda|}.
\end{aligned}$$

This proves that the operator $A + \delta I$ satisfies condition (1.1), so that estimates (1.4) and (1.5) remain valid for the operator $A + \delta I$.

$$\|V(z)\| \leq \widetilde{M_0}(\varepsilon), \quad z \in \Delta_\omega^{2\varepsilon}, \tag{1.14}$$

$$\|(A + \delta I)V(z)\| \leq \frac{\widetilde{M_1}(\varepsilon)}{|z|}, \quad z \in \Delta_\omega^{2\varepsilon}, \tag{1.15}$$

where

$$V(z) = -\frac{1}{2\pi i} \int_\Gamma e^{\lambda z}(A + \delta I - \lambda I)^{-1}\, d\lambda.$$

But we have by Cauchy's theorem

$$\begin{aligned}
V(z) &= -\frac{1}{2\pi i} \int_\Gamma e^{\lambda z}(A + \delta I - \lambda I)^{-1}\, d\lambda \\
&= -\frac{1}{2\pi i} \int_{\Gamma + \delta} e^{\lambda z}(A + \delta I - \lambda I)^{-1}\, d\lambda \\
&= -\frac{1}{2\pi i} \int_\Gamma e^{\mu z} e^{\delta z}(A - \mu I)^{-1}\, d\mu = e^{\delta z} U(z).
\end{aligned} \tag{1.16}$$

In view of formula (1.16), the desired estimates (1.4′) and (1.5′) follow from estimates (1.14) and (1.15). □

1.2 Fractional Powers

Assume that the operator A satisfies a stronger condition than condition (1.1).

(1) The resolvent set of A contains the region Σ shown in Figure 1.8.

(2) There exists a constant $M > 0$ such that the resolvent $R(\lambda) = (A - \lambda I)^{-1}$ satisfies the estimate

$$\|R(\lambda)\| \leq \frac{M}{1+|\lambda|}, \quad \lambda \in \Sigma. \tag{1.17}$$

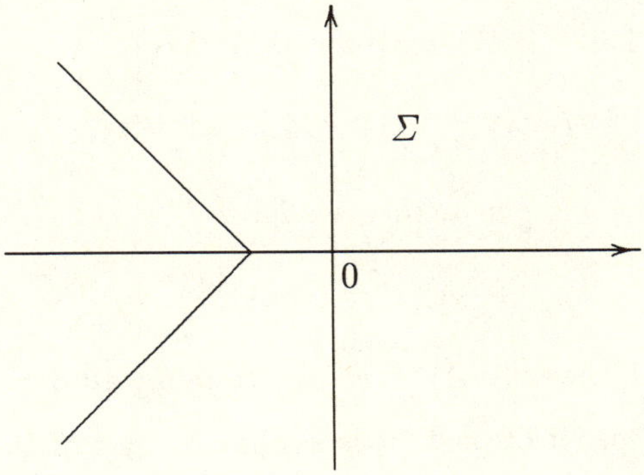

Figure 1.8

If $\alpha > 0$, we can define the fractional power $(-A)^{-\alpha}$ of $-A$ by the following formula:

$$(-A)^{-\alpha} = -\frac{1}{2\pi i}\int_\Gamma (-\lambda)^{-\alpha} R(\lambda)\, d\lambda. \tag{1.18}$$

Here the path Γ runs in the set Σ from $\infty e^{-i\omega}$ to $\infty e^{i\omega}$, avoiding the positive real axis and the origin (see Figure 1.9), and for the function $(-\lambda)^{-\alpha} = e^{-\alpha \log(-\lambda)}$, we choose the branch whose argument lies between $-\alpha\pi$ and $\alpha\pi$; it is analytic in the region obtained by omitting the positive real axis.

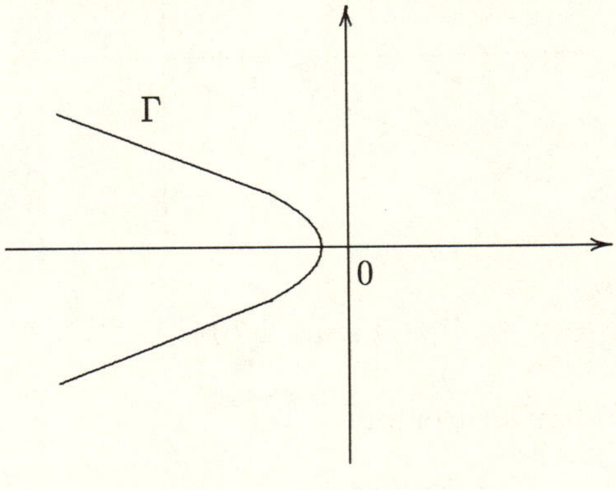

Figure 1.9

The integral (1.18) converges in the uniform operator topology for $\alpha > 0$, and thus defines a bounded linear operator on E. Indeed, it suffices to note the following:

$$\begin{cases} |(-\lambda)^{-\alpha}| = \left|e^{-\alpha \log(-\lambda)}\right| = e^{-\alpha \log|\lambda|} = |\lambda|^{-\alpha}, \\ \|R(\lambda)\| \leq \frac{M}{1+|\lambda|}. \end{cases}$$

Some basic properties of the fractional power $(-A)^{-\alpha}$ are summarized in the following:

Proposition 1.4. *(i) We have for all $\alpha, \beta > 0$*

$$(-A)^{-\alpha}(-A)^{-\beta} = (-A)^{-(\alpha+\beta)}.$$

(ii) If α is a positive integer n, then we have

$$(-A)^{-\alpha} = \left((-A)^{-1}\right)^n.$$

(iii) The fractional power $(-A)^{-\alpha}$ is invertible for all $\alpha > 0$.

Proof. (i) The proof of part (i) is similar to that of Proposition 1.1. By Cauchy's theorem, one may assume that

$$(-A)^{-\beta} = -\frac{1}{2\pi i}\int_{\Gamma'} (-\mu)^{-\beta} R(\mu)\, d\mu,$$

where Γ' is a path obtained from the path Γ by translating each point of Γ to the right by a fixed small positive distance (see Figure 1.10).

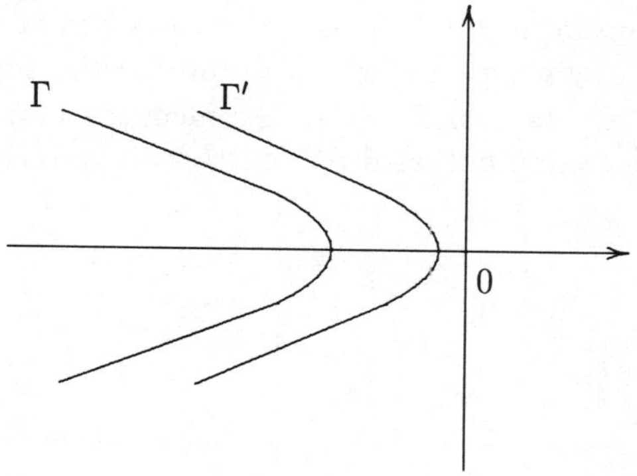

Figure 1.10

Then we have by Fubini's theorem

$$(-A)^{-\alpha}(-A)^{-\beta} = \frac{1}{(2\pi i)^2}\int_{\Gamma}\int_{\Gamma'} (-\lambda)^{-\alpha}(-\mu)^{-\beta} R(\lambda)\, R(\mu)\, d\lambda\, d\mu$$

$$= \frac{1}{(2\pi i)^2} \int_{\Gamma} \int_{\Gamma'} (-\lambda)^{-\alpha} (-\mu)^{-\beta} \frac{R(\lambda) - R(\mu)}{\lambda - \mu} \, d\lambda \, d\mu$$
$$= \frac{1}{2\pi i} \int_{\Gamma} (-\lambda)^{-\alpha} R(\lambda) \left[\frac{1}{2\pi i} \int_{\Gamma'} \frac{(-\mu)^{-\beta}}{\lambda - \mu} \, d\mu \right] d\lambda$$
$$- \frac{1}{2\pi i} \int_{\Gamma'} (-\mu)^{-\beta} R(\mu) \left[\frac{1}{2\pi i} \int_{\Gamma} \frac{(-\lambda)^{-\alpha}}{\lambda - \mu} \, d\lambda \right] d\mu.$$

Just as in the proof of Proposition 1.1, we can calculate each term in the last part as follows:

$$\frac{1}{2\pi i} \int_{\Gamma'} \frac{(-\mu)^{-\beta}}{\lambda - \mu} \, d\mu = -(-\lambda)^{-\beta};$$
$$\frac{1}{2\pi i} \int_{\Gamma} \frac{(-\lambda)^{-\alpha}}{\lambda - \mu} \, d\lambda = 0.$$

Therefore, we obtain that

$$(-A)^{-\alpha}(-A)^{-\beta} = -\frac{1}{2\pi i} \int_{\Gamma} (-\lambda)^{-(\alpha+\beta)} R(\lambda) \, d\lambda = (-A)^{-(\alpha+\beta)}.$$

(ii) Since we have by estimate (1.17)

$$\lim_{r \to \infty} \int_{-\omega}^{\omega} (-re^{i\theta})^{-n} R(re^{i\theta}) ire^{i\theta} \, d\theta = 0 \quad \text{for all integer } n \geq 1,$$

it follows that

$$(-A)^{-n} = -\frac{1}{2\pi i} \int_{\Gamma} (-\lambda)^{-n} R(\lambda) \, d\lambda = \frac{1}{2\pi i} \lim_{r \to \infty} \int_{C_r} (-\lambda)^{-n} R(\lambda) \, d\lambda,$$

where C_r is a closed path shown in Figure 1.11.

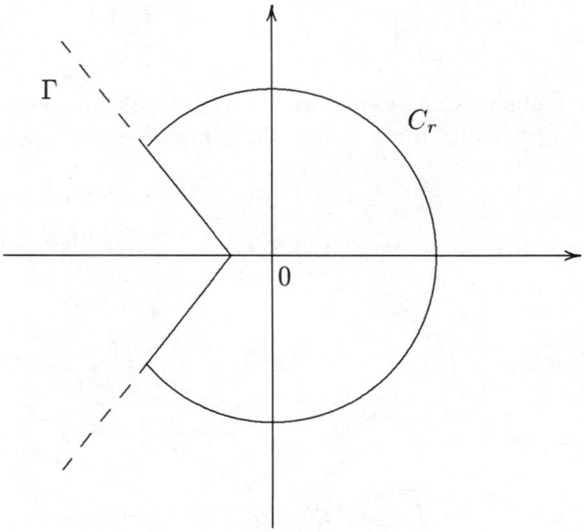

Figure 1.11

Thus, by the residue theorem, we obtain that

$$\begin{aligned}(-A)^{-n} &= \operatorname{Res}\left[(-\lambda)^{-n} R(\lambda)\right]_{\lambda=0} \\ &= \frac{(-1)^n}{(n-1)!} \frac{d^{n-1}}{d\lambda^{n-1}} \left((A - \lambda I)^{-1}\right)\bigg|_{\lambda=0} \\ &= (-1)^n (A^{-1})^n \\ &= ((-A)^{-1})^n.\end{aligned}$$

(iii) Since the operator $(-A)^{-1}$ is injective, it follows that $(-A)^{-n} = ((-A)^{-1})^n$ is injective for all integer $n \geq 1$. Assume that

$$(-A)^{-\alpha} x = 0, \quad \alpha > 0.$$

Then, taking an integer $n > \alpha$, we obtain that

$$(-A)^{-n} x = (-A)^{-(n-\alpha)} (-A)^{-\alpha} x = 0,$$

so that

$$x = 0.$$

This proves part (iii).

The proof of Proposition 1.4 is complete. □

If $0 < \alpha < 1$, we have the following useful formula for the fractional power $(-A)^{-\alpha}$:

Theorem 1.5. *We have for* $0 < \alpha < 1$

$$(1.19) \qquad (-A)^{-\alpha} = -\frac{\sin \alpha\pi}{\pi} \int_0^\infty s^{-\alpha} R(s)\, ds.$$

Proof. By Cauchy's theorem, one may deform the path Γ in formula (1.18) into the upper and lower sides of the positive real axis. But we have

$$(-\lambda)^{-\alpha} = e^{-\alpha \log(-\lambda)} = \begin{cases} |\lambda|^{-\alpha} e^{\alpha\pi i} & \text{if } \operatorname{Im} \lambda > 0, \\ |\lambda|^{-\alpha} e^{-\alpha\pi i} & \text{if } \operatorname{Im} \lambda < 0. \end{cases}$$

Hence it follows that

$$\begin{aligned}(-A)^{-\alpha} &= -\frac{1}{2\pi i} \int_0^\infty s^{-\alpha} e^{\alpha\pi i} R(s)\, ds - \frac{1}{2\pi i} \int_\infty^0 s^{-\alpha} e^{-\alpha\pi i} R(s)\, ds \\ &= -\frac{1}{\pi} \frac{e^{\alpha\pi i} - e^{\alpha\pi i}}{2i} \int_0^\infty s^{-\alpha} R(s)\, ds \\ &= -\frac{\sin \alpha\pi}{\pi} \int_0^\infty s^{-\alpha} R(s)\, ds. \quad \square\end{aligned}$$

Corollary 1.6. *We have for* $0 < \alpha < 1$

$$\|(-A)^{-\alpha}\| \leq M. \tag{1.20}$$

Proof. By formula (1.19), it follows that

$$\begin{aligned}\|(-A)^{-\alpha}\| &\leq \frac{\sin \alpha\pi}{\pi} \int_0^\infty s^{-\alpha} \|R(s)\|\, ds \\ &\leq M \frac{\sin \alpha\pi}{\pi} \int_0^\infty s^{-\alpha}(1+s)^{-1}\, ds.\end{aligned}$$

But we have

$$\begin{aligned}\int_0^\infty s^{-\alpha}(1+s)^{-1}\, ds &= \int_0^1 \left(\frac{\sigma}{1-\sigma}\right)^{-\alpha} (1-\sigma)\frac{d\sigma}{(1-\sigma)^2} \\ &= \int_0^1 \sigma^{-\alpha}(1-\sigma)^{\alpha-1}\, d\sigma \\ &= B(1-\alpha, \alpha) \\ &= \Gamma(1-\alpha)\Gamma(\alpha),\end{aligned}$$

and also the well-known formula

$$\frac{\sin \alpha\pi}{\pi} = \frac{1}{\Gamma(\alpha)\Gamma(1-\alpha)}.$$

Summing up, we obtain that

$$\|(-A)^{-\alpha}\| \leq M. \quad \square$$

By Remark 1.3, one may assume that there exist positive constants M_0, M_1 and a such that

$$\|U(t)\| \leq M_0 e^{-at}, \quad t \geq 0, \tag{1.21}$$

$$\|AU(t)\| \leq M_1 e^{-at}\frac{1}{t}, \quad t > 0. \tag{1.22}$$

Then we can prove still another useful formula for the fractional power $(-A)^{-\alpha}$, $0 < \alpha < 1$.

To this end, we need a representation formula for the resolvent $R(s) = (A - sI)^{-1}$, $s \geq 0$.

Lemma 1.7. *We have for all $s \geq 0$*

$$R(s) = -\int_0^\infty e^{-st} U(t)\, dt. \tag{1.23}$$

Proof. Let $s > 0$. For any $T > 0$, we have by Fubini's theorem

$$\begin{aligned}
\int_0^T e^{-st} U(t)\, dt &= -\frac{1}{2\pi i} \int_0^T e^{-st} \left(\int_\Gamma e^{\mu t} R(\mu)\, d\mu \right) dt \\
&= -\frac{1}{2\pi i} \int_\Gamma \left(\int_0^T e^{\mu t - st} dt \right) R(\mu) d\mu \\
&= -\frac{1}{2\pi i} \int_\Gamma \frac{1}{\mu - s} \left(e^{(\mu - s)T} - 1 \right) R(\mu) d\mu \\
&= \frac{1}{2\pi i} \int_\Gamma \frac{1}{\mu - s} R(\mu) d\mu - \frac{1}{2\pi i} \int_\Gamma e^{(\mu - s)T} \frac{R(\mu)}{\mu - s}\, d\mu \\
&= -R(s) - \frac{1}{2\pi i} \int_\Gamma e^{(\mu - s)T} \frac{R(\mu)}{\mu - s}\, d\mu.
\end{aligned} \tag{1.24}$$

But we find that the second term in the last line tends to zero as $T \to \infty$:

$$\left\| \int_\Gamma e^{(\mu - s)T} \frac{R(\mu)}{\mu - s}\, d\mu \right\| \leq e^{-sT} \int_\Gamma \frac{M}{1 + |\mu|} \frac{|\, d\mu|}{|\mu - s|} \longrightarrow 0.$$

Thus formula (1.23) for $s > 0$ follows by letting $T \to \infty$ in formula (1.24).

In view of estimate (1.21), it follows that the integral $\int_0^\infty e^{-st} U(t) dt$ converges uniformly in $s \in [0, \infty)$ in the space $L(E, E)$. Hence, applying the Lebesgue dominated convergence theorem, we obtain that formula (1.23) remains valid for $s = 0$. $\square$

Theorem 1.8. *We have for $0 < \alpha < 1$*

$$(-A)^{-\alpha} = \frac{1}{\Gamma(\alpha)} \int_0^\infty t^{\alpha - 1} U(t)\, dt. \tag{1.19$'$}$$

Proof. Substituting formula (1.23) into formula (1.19), we obtain that

$$\begin{aligned}
(-A)^{-\alpha} &= \frac{\sin \alpha\pi}{\pi} \int_0^\infty s^{-\alpha} \left(\int_0^\infty e^{-st} U(t) dt \right) ds \\
&= \frac{\sin \alpha\pi}{\pi} \int_0^\infty U(t) \left(\int_0^\infty e^{-st} s^{-\alpha}\, ds \right) dt \\
&= \frac{\sin \alpha\pi}{\pi} \int_0^\infty U(t) \left(\int_0^\infty e^{-\tau} \left(\frac{\tau}{t} \right)^{-\alpha} \frac{d\tau}{t} \right) dt
\end{aligned}$$

$$= \frac{\sin \alpha\pi}{\pi} \int_0^\infty e^{-\tau} \tau^{-\alpha} \, d\tau \cdot \int_0^\infty t^{\alpha-1} U(t) \, dt$$
$$= \frac{1}{\Gamma(\alpha)} \int_0^\infty t^{\alpha-1} U(t) \, dt. \quad \square$$

In view of part (iii) of Proposition 1.4, we can define the fractional power $(-A)^\alpha$ for $\alpha > 0$ as follows:

$$(-A)^\alpha = \text{the inverse of } (-A)^{-\alpha}, \quad \alpha > 0.$$

The next theorem states that the domain $D((-A)^\alpha)$ of $(-A)^\alpha$ is bigger than the domain $D(A)$ of A when $0 < \alpha < 1$.

Theorem 1.9. *We have for any* $0 < \alpha < 1$

$$D(A) \subset D((-A)^\alpha).$$

Proof. Let x be an arbitrary element of $D(A)$. Then there exists a unique element $y \in E$ such that

$$x = (-A)^{-1} y.$$

If $0 < \alpha < 1$, one can define the fractional powers $(-A)^{-\alpha}$ and $(-A)^{-(1-\alpha)}$, and write $(-A)^{-1}$ as follows:

$$(-A)^{-1} = (-A)^{-\alpha} (-A)^{-(1-\alpha)}.$$

Hence we have

$$x = (-A)^{-1} y = (-A)^{-\alpha} \left((-A)^{-(1-\alpha)} y \right).$$

This proves that

$$x \in D((-A)^\alpha).$$

The proof of Theorem 1.9 is complete. $\square$

We can give an explicit formula for the fractional power $(-A)^\alpha$, $0 < \alpha < 1$, on the domain $D(A)$.

Theorem 1.10. *Let* $0 < \alpha < 1$. *We have for any* $x \in D(A)$

$$(-A)^\alpha x = \frac{\sin \alpha\pi}{\pi} \int_0^\infty s^{\alpha-1} R(s) A x \, ds. \tag{1.25}$$

Proof. First we remark that

$$(-A)^\alpha = (-A)(-A)^{-(1-\alpha)}.$$

By formula (1.19) with $1-\alpha$ for α, it follows that

$$\begin{aligned}(-A)^{-(1-\alpha)} &= -\frac{\sin(1-\alpha)\pi}{\pi}\int_0^\infty s^{\alpha-1}R(s)\,ds \\ &= -\frac{\sin\alpha\pi}{\pi}\int_0^\infty s^{\alpha-1}R(s)\,ds.\end{aligned}$$

But we have for $x \in D(A)$

$$|s^{\alpha-1}R(s)Ax| \leq \begin{cases} s^{\alpha-1}(1+M)\|x\| & \text{near } s=0, \\ s^{\alpha-2}M\|Ax\| & \text{near } s=\infty, \end{cases} \tag{1.26}$$

since $R(s)A = I + sR(s)$ and $|R(s)| \leq M/s$, $s>0$. This implies that

$$\int_0^\infty s^{\alpha-1}\|R(s)Ax\|\,ds < \infty.$$

Hence, by the closedness of A, it follows that

$$(-A)^{-(1-\alpha)}x \in D(A)$$

and

$$\begin{aligned}(-A)^\alpha x &= (-A)(-A)^{-(1-\alpha)}x \\ &= \frac{\sin\alpha\pi}{\pi}\int_0^\infty s^{\alpha-1}R(s)Ax\,ds. \quad \square\end{aligned}$$

Corollary 1.11 (The moment inequality). *Let $0<\alpha<1$. Then we have*

$$\|(-A)^\alpha x\| \leq 2M(1+M)\|x\|^{1-\alpha}\|Ax\|^\alpha, \quad x \in D(A). \tag{1.27}$$

Proof. By formula (1.25) and estimate (1.26), it follows that for every $\rho > 0$

$$\begin{aligned}\|(-A)^\alpha x\| &= \left\|\frac{\sin\alpha\pi}{\pi}\int_0^\infty s^{\alpha-1}R(s)Ax\,ds\right\| \\ &\leq \left|\frac{\sin\alpha\pi}{\alpha\pi}\right|\int_0^\rho \alpha s^{\alpha-1}\|R(s)Ax\|\,ds \\ &\quad + \left|\frac{\sin(1-\alpha)\pi}{(1-\alpha)\pi}\right|\int_\rho^\infty (1-\alpha)s^{\alpha-1}\|R(s)Ax\|\,ds \\ &\leq \left|\frac{\sin\alpha\pi}{\alpha\pi}\right|\int_0^\rho \alpha s^{\alpha-1}(1+M)\,ds\cdot\|x\|\end{aligned}$$

$$+\left|\frac{\sin(1-\alpha)\pi}{(1-\alpha)\pi}\right|\int_\rho^\infty (1-\alpha)s^{\alpha-2}M\,ds\cdot\|Ax\|$$
$$\leq (1+M)\|x\|\rho^\alpha + M\|Ax\|\rho^{\alpha-1}.$$

Therefore, the moment inequality (1.27) follows by taking

$$\rho = \frac{M}{1+M}\frac{\|Ax\|}{\|x\|}.$$

The proof of Corollary 1.11 is complete. □

Now we study some useful relationships between the fractional powers $(-A)^\alpha$, $0<\alpha\leq 1$, and the semigroup $U(t)$.

Theorem 1.12. *Let $0<\alpha\leq 1$. For all $t>0$, we have the following:*
(a) $U(t): E \longrightarrow D((-A)^\alpha)$.
(b) $U(t)(-A)^\alpha x = (-A)^\alpha U(t)x$, $x\in D((-A)^\alpha)$.
(c) $\|(-A)^\alpha U(t)\| \leq 2M(1+M)M_0^{1-\alpha}M_1^\alpha t^{-\alpha}e^{-at}$.
(d) $\|U(t)x - x\| \leq \frac{2M(1+M)}{\alpha}M_0^\alpha M_1^{1-\alpha}t^\alpha\|(-A)^\alpha x\|$, $x\in D((-A)^\alpha)$.

Proof. (a) This is obvious, since $U(t): E\to D(A)$ for every $t>0$ and $D(A)\subset D((-A)^\alpha)$.

(b) If $x\in D((-A)^\alpha)$, then we have by formula (1.19′)

$$\begin{aligned}
U(t)x &= U(t)((-A)^{-\alpha})(-A)^\alpha x\\
&= \frac{1}{\Gamma(\alpha)}\int_0^\infty s^{\alpha-1}U(t)(U(s)(-A)^\alpha x)\,ds\\
&= \frac{1}{\Gamma(\alpha)}\int_0^\infty s^{\alpha-1}U(s)(U(t)(-A)^\alpha x)\,ds\\
&= (-A)^{-\alpha}U(t)(-A)^\alpha x.
\end{aligned}$$

This proves part (b).

(c) Since $(-A)^\alpha = ((-A)^{-\alpha})^{-1}$ is closed and $U(t)$ is bounded, it follows that the operator $(-A)^\alpha U(t)$ is closed. But part (a) tells us that it is everywhere defined on E. Hence, by the closed graph theorem, it follows that $(-A)^\alpha U(t)$ is a bounded operator on E. More precisely, applying the moment inequality (1.27) to $U(t)x$, we obtain from estimates (1.21) and (1.22) that

$$\begin{aligned}
\|(-A)^\alpha U(t)x\| &\leq 2M(1+M)\|U(t)x\|^{1-\alpha}\|AU(t)x\|^\alpha\\
&\leq 2M(1+M)\left(M_0e^{-at}\|x\|\right)^{1-\alpha}\left(M_1e^{-at}t^{-1}\|x\|\right)^\alpha\\
&\leq 2M(1+M)M_0^{1-\alpha}M_1^\alpha t^{-\alpha}e^{-at}\|x\|.
\end{aligned}$$

This proves part (c).

(d) By part (b), we have for all $x \in D((-A)^\alpha)$

$$\begin{aligned} U(t)x - x &= \int_0^t U'(s)x\,ds \\ &= \int_0^t AU(s)x\,ds \\ &= \int_0^t (-A)^{1-\alpha}U(s)(-A)^\alpha x\,ds. \end{aligned}$$

Hence it follows from part (c) with $1-\alpha$ for α that

$$\begin{aligned} \|U(t)x - x\| &\le \int_0^t \|(-A)^{1-\alpha}U(s)\| \cdot \|(-A)^\alpha x\|\,ds \\ &\le 2M(1+M)M_0^\alpha M_1^{1-\alpha} \int_0^t s^{\alpha-1}\,ds \cdot \|(-A)^\alpha x\| \\ &\le \frac{2M(1+M)}{\alpha} M_0^\alpha M_1^{1-\alpha} t^\alpha \|(-A)^\alpha x\|. \end{aligned}$$

The proof of Theorem 1.12 is complete. □

The next lemma is useful in applications to concrete initial boundary value problems for semilinear parabolic differential equations (cf. [He, Section 1.4, Exercise 11]).

Lemma 1.13. *Let $A : E \to E$ be a densely defined, closed linear operator which satisfies condition (1.17), and let B be a closed linear operator from $D(B) \subset E$ into a Banach space F. Assume that $D(B) \supset D(A)$, and that there exist constants $0 < \gamma < 1$ and $\delta_0 > 0$ such that, for all $0 < \delta \le \delta_0$,*

$$\|Bx\|_F \le C\left(\delta^{-\gamma}\|x\| + \delta^{1-\gamma}\|Ax\|\right), \quad x \in D(A). \tag{1.28}$$

Then we have for all $\gamma < \alpha < 1$

$$D((-A)^\alpha) \subset D(B),$$

and

$$\|Bz\|_F \le K_\alpha \|(-A)^\alpha z\|, \quad z \in D((-A)^\alpha),$$

with a constant $K_\alpha > 0$.

Proof. (1) We show that: If $(-A)^{-\alpha}x \in D(B)$, then we have

$$B((-A)^{-\alpha}x) = \frac{1}{\Gamma(\alpha)} \int_0^\infty s^{\alpha-1} BU(s)x\,ds. \tag{1.29}$$

We use the representation formula (1.19′) for the fractional power $(-A)^{-\alpha}$

$$(-A)^{-\alpha} = \frac{1}{\Gamma(\alpha)} \int_0^\infty s^{\alpha-1} U(s)\,ds. \tag{1.19′}$$

Since B is closed, it suffices to show that the integral $\int_0^\infty s^{\alpha-1} BU(s)x \, ds$ is convergent in F.

First we have

$$\begin{aligned}
&\left\| \int_0^\infty s^{\alpha-1} BU(s)x \, ds \right\|_F \\
&\leq \int_0^\infty s^{\alpha-1} \|BU(s)x\|_F \, ds \\
&\leq \int_0^{\delta_0} s^{\alpha-1} \|BU(s)x\|_F \, ds + \int_{\delta_0}^\infty s^{\alpha-1} \|BU(s)x\|_F \, ds.
\end{aligned}$$

We estimate each term in the last line.

(1) Applying inequality (1.28) with $U(s)x$ for x and s for δ, we obtain from estimates (1.21) and (1.22) that

$$\begin{aligned}
\|BU(s)x\|_F &\leq C \left(s^{-\gamma} \|U(s)x\| + s^{1-\gamma} \|AU(s)x\| \right) \\
&\leq C \left(M_0 s^{-\gamma} + s^{1-\gamma} M_1 s^{-1} \right) \|x\| \\
&= C(M_0 + M_1) s^{-\gamma} \|x\|.
\end{aligned}$$

Hence we have for $\gamma < \alpha < 1$

$$\begin{aligned}
\int_0^{\delta_0} s^{\alpha-1} \|BU(s)x\|_F \, ds &\leq C(M_0 + M_1) \left(\int_0^{\delta_0} s^{\alpha-\gamma-1} \, ds \right) \|x\| \\
&= \frac{C(M_0 + M_1)}{\alpha - \gamma} \delta_0^{\alpha-\gamma} \|x\|.
\end{aligned} \tag{1.30}$$

(2) Similarly, applying inequality (1.28) with $x = U(s)x$ and $\delta = \delta_0$, we obtain that

$$\begin{aligned}
&\int_{\delta_0}^\infty s^{\alpha-1} \|BU(s)x\|_F \, ds \\
&\leq C \int_{\delta_0}^\infty s^{\alpha-1} \left(\delta_0^{-\gamma} \|U(s)x\| + \delta_0^{1-\gamma} \|AU(s)x\| \right) ds \\
&\leq C \int_{\delta_0}^\infty s^{\alpha-1} \left(\delta_0^{-\gamma} M_0 + \delta_0^{1-\gamma} M_1 s^{-1} \right) e^{-as} \, ds \cdot \|x\| \\
&\leq C \left(M_0 \delta_0^{\alpha-\gamma-1} \int_0^\infty e^{-as} \, ds + M_1 \delta_0^{1-\gamma} \int_{\delta_0}^\infty s^{\alpha-2} e^{-as} \, ds \right) \|x\| \\
&\leq C \left(M_0 \delta_0^{\alpha-\gamma-1} \frac{1}{a} + M_1 \delta_0^{1-\gamma} \left(\left[-\frac{1}{a} e^{-as} s^{\alpha-2} \right]_{\delta_0}^\infty \right.\right. \\
&\quad \left.\left. + \frac{1}{a}(2-\alpha) \int_{\delta_0}^\infty s^{\alpha-3} \, ds \right) \right) \|x\| \\
&\leq \frac{C}{a} \left(M_0 + 2M_1 \right) \delta_0^{\alpha-\gamma-1} \|x\|.
\end{aligned} \tag{1.31}$$

Therefore, combining inequalities (1.30) and (1.31), we have

$$\frac{1}{\Gamma(\alpha)} \int_0^\infty s^{\alpha-1} \|BU(s)x\|_F \, ds \leq K_\alpha \|x\|,$$

with

$$K_\alpha = \frac{C}{\Gamma(\alpha)} \left(\frac{1}{\alpha - \gamma} (M_0 + M_1) \delta_0^{\alpha-\gamma} + \frac{1}{a} (M_0 + 2M_1) \delta_0^{\alpha-\gamma-1} \right).$$

This proves formula (1.29) and also

$$\|B((-A)^{\alpha})x\|_F \leq K_\alpha \|x\|, \quad x \in D(B(-A)^{-\alpha}). \tag{1.32}$$

(2) We remark that

$$(-A)^{1-\alpha} = (-A)(-A)^{-\alpha}.$$

Hence we have for all $x \in D((-A)^{1-\alpha})$

$$(-A)^{-\alpha} x \in D(A) \subset D(B).$$

Thus we can apply inequality (1.32) to obtain that

$$\|B((-A)^{-\alpha})x\|_F \leq K_\alpha \|x\|, \quad x \in D((-A)^{1-\alpha}). \tag{1.32$'$}$$

Now let x be an arbitrary element of E. Since the domain $D(A)$ is dense in E and since $D(A) \subset D((-A)^{1-\alpha})$, we can choose a sequence $\{x_j\}$ in $D((-A)^{1-\alpha})$ such that

$$x_j \longrightarrow x \quad \text{in } E. \tag{1.33}$$

Then, by inequality (1.32′), it follows that $\{B((-A)^{-\alpha})x_j\}$ is a Cauchy sequence in F. Hence there exists an element y of F such that

$$B((-A)^{-\alpha})x_j \to y \quad \text{in } F. \tag{1.34}$$

But, since B is closed and $(-A)^{-\alpha}$ is bounded, it follows that $B((-A)^{-\alpha})$ is a closed operator. Hence we have by (1.33) and (1.34)

$$\begin{cases} x \in D(B((-A)^{-\alpha})), \\ B((-A)^{-\alpha} x) = y. \end{cases}$$

This proves that

$$D\left(B((-A)^{-\alpha})\right) = E,$$

so that

$$D((-A)^{\alpha}) \subset D(B).$$

Finally, we can pass to the limit in the inequality

$$\left\|B\left((-A)^{-\alpha}\right) x_j\right\|_F \leq K_\alpha \|x_j\| \tag{1.32$'$}$$

to obtain that

$$\left\|B\left((-A)^{-\alpha}\right) x\right\|_F \leq K_\alpha \|x\|, \quad x \in E,$$

or equivalently (letting $z = (-A)^{-\alpha} x$)

$$\|Bz\|_F \leq K_\alpha \, \|(-A)^{\alpha} z\|, \quad z \in D((-A)^{\alpha}).$$

The proof of Lemma 1.13 is complete. □

1.3 The Linear Cauchy Problem

In this section, we consider the following Cauchy problem:

$$\text{(P)} \qquad \begin{cases} \frac{dx}{dt} = Ax(t), & 0 < t < T, \\ x(0) = x_0. \end{cases}$$

A function $x(t) : [0,T) \to E$ is called a *solution* of problem (P) if it satisfies the following three conditions:

(1) $x(t) \in C([0,T); E) \cap C^1((0,T); E)$ and $x(0) = x_0$.
(2) $x(t) \in D(A)$ for all $0 < t < T$.
(3) $\frac{dx}{dt} = Ax(t)$ for all $0 < t < T$.

Here $C([0,T); E)$ denotes the space of continuous functions on $[0,T)$ taking values in E, and $C^1((0,T); E)$ denotes the space of continuously differentiable functions on $(0,T)$ taking values in E, respectively.

Assume that the operator A satisfies condition (1.17):

$$\text{(1.17)} \qquad \|R(\lambda)\| \le \frac{M}{1+|\lambda|}, \quad \lambda \in \Sigma.$$

The next theorem states that problem (P) has a unique solution $x(t)$ for any initial condition $x_0 \in E$.

Theorem 1.14. *Assume that the operator A satisfies condition (1.17). Then, for every $x_0 \in E$, the function*

$$\text{(1.35)} \qquad x(t) = U(t)x_0 = -\frac{1}{2\pi i}\int_\Gamma e^{\lambda t} R(\lambda) x_0 \, d\lambda, \quad t > 0,$$

belongs to the space $C([0,\infty); E) \cap C^\infty((0,\infty); E)$, and is a unique solution of problem (P).

Proof. By Theorem 1.2, it suffices to prove that the function $U(t)x_0$ is the only solution of problem (P), for every $T > 0$.

Let $y(t)$ be an arbitrary solution of problem (P), and let

$$z(t,s) = U(t-s)y(s), \quad 0 \le s \le t < T.$$

Then we obtain that the function

$$s \longmapsto z(t,s)$$

belongs to the space $C([0,t]; E) \cap C^1((0,t); E)$, and satisfies

$$\begin{aligned} \frac{\partial}{\partial s}(z(t,s)) &= U(t-s)y'(s) - AU(t-s)y(s) \\ &= U(t-s)(y'(s) - Ay(s)) \end{aligned}$$

$$= 0, \quad 0 < s < t.$$

This implies that

$$z(t,0) = z(t,t),$$

so that

$$y(t) = U(t)x_0, \quad 0 \le t < T.$$

This proves Theorem 1.14. □

Let $f : [0,T) \to E$ be a continuous function. Now we consider the following non-homogeneous Cauchy problem:

$$\text{(NP)} \qquad \begin{cases} \frac{dx}{dt} = Ax(t) + f(t), & 0 < t < T, \\ x(0) = x_0. \end{cases}$$

The next theorem gives an explicit formula for the solutions of problem (NP).

Theorem 1.15. *Assume that the operator A satisfies condition (1.17). Then a solution $x(t)$ of problem (NP), if it exists, is given by the following formula:*

$$x(t) = U(t)x_0 + \int_0^t U(t-s)f(s)\,ds, \quad 0 < t < T. \tag{1.36}$$

Proof. Applying the operators $U(t-s)$, $0 < s < t$, to the equation

$$x'(s) = Ax(s) + f(s), \quad 0 < s < T,$$

we obtain that

$$U(t-s)x'(s) = U(t-s)Ax(s) + U(t-s)f(s), \quad 0 < s < t. \tag{1.37}$$

On the other hand, it follows that

$$\begin{aligned} \frac{d}{ds}(U(t-s)x(s)) &= \lim_{\sigma \to 0} \left(\frac{U(t-s-\sigma)x(s+\sigma) - U(t-s)x(s)}{\sigma} \right) \\ &= \lim_{\sigma \to 0} \left\{ U(t-s-\sigma) \left(\frac{x(s+\sigma) - x(s)}{\sigma} \right) \right. \\ &\qquad \left. - U(t-s-\sigma) \left(\frac{U(\sigma) - I}{\sigma} \right) x(s) \right\} \\ &= U(t-s)x'(s) - U(t-s)Ax(s), \quad 0 < s < t. \end{aligned} \tag{1.38}$$

Hence it follows from formulas (1.37) and (1.38) that

$$\frac{d}{ds}(U(t-s)x(s)) = U(t-s)f(s), \quad 0 < s < t < T.$$

Integrating this equation from 0 to $t-h$, $h>0$, with respect to s, we obtain that

$$\begin{aligned}\int_0^{t-h} U(t-s)f(s)\,ds &= [U(t-s)x(s)]_0^{t-h} \\ &= U(h)x(t-h) - U(t)x_0.\end{aligned}$$

But we have, as $h \downarrow 0$,

$$\begin{aligned}\|U(h)x(t-h) - x(t)\| \leq &\|U(h)\| \cdot \|x(t-h) - x(t)\| + \|(U(h) - I)x(t)\| \\ &\longrightarrow 0.\end{aligned}$$

Indeed, it suffices to note the following:

(1) $\sup_{0<h\leq T} \|U(h)\| < \infty$.
(2) $\lim_{h\downarrow 0} U(h)x = x$ for each $x \in E$.
(3) $x(t) \in C([0,T);E)$.

Therefore, we find that the improper integral $\int_0^t U(t-s)f(s)\,ds$ exists, and satisfies

$$\int_0^t U(t-s)f(s)\,ds = x(t) - U(t)x_0.$$

This proves formula (1.36).

The proof of Theorem 1.15 is complete. □

The next theorem states that the function $x(t)$, defined by formula (1.36), is a solution of problem (NP) (cf. [He, Theorem 3.2.2]; [Pa, Chapter 4, Corollary 3.3]).

Theorem 1.16. *Assume that the operator A satisfies condition (1.17). Let f be a locally Hölder continuous function on $(0,T)$, with exponent $0 < \gamma \leq 1$, which satisfies the condition*

$$\int_0^T \|f(s)\|\,ds < \infty. \tag{1.39}$$

Then, for every $x_0 \in E$, the function

$$x(t) = U(t)x_0 + \int_0^t U(t-s)f(s)\,ds \tag{1.36}$$

belongs to the space $C([0,T];E) \cap C^1((0,T);E)$, and is a unique solution of problem (NP).

Proof. Theorem 1.14 tells us that the function $U(t)x_0$, given by formula (1.35), belongs to $C([0,\infty);E) \cap C^\infty((0,\infty);E)$, and is a (unique) solution of problem (P).

Thus it suffices to consider the function

$$y(t) = \int_0^t U(t-s)f(s)\,ds, \quad 0 < t < T. \tag{1.36$'$}$$

First we have by estimate (1.21)

$$\|y(t)\| \leq \int_0^t \|U(t-s)\| \cdot \|f(s)\|\,ds \leq M_0 \int_0^t \|f(s)\|\,ds,$$

so that

$$\lim_{t \downarrow 0} y(t) = 0.$$

(a) The continuity of $y(t)$: For each small $\delta > 0$, we let

$$y_\delta(t) = \begin{cases} 0 & \text{if } 0 \leq t \leq \delta, \\ \int_0^{t-\delta} U(t-s)f(s)\,ds & \text{if } \delta < t < T. \end{cases}$$

Then we have

$$\begin{aligned} \|y_\delta(t) - y(t)\| &\leq \begin{cases} \int_0^t \|U(t-s)\| \cdot \|f(s)\|\,ds & \text{if } 0 \leq t \leq \delta, \\ \int_{t-\delta}^t \|U(t-s)\| \cdot \|f(s)\|\,ds & \text{if } \delta < t < T. \end{cases} \\ &\leq \int_0^\delta \|U(t-s)\| \cdot \|f(s)\|\,ds \\ &\quad + \int_{t-\delta}^t \|U(t-s)\| \cdot \|f(s)\|\,ds \\ &\leq M_0 \left(\int_0^\delta \|f(s)\|\,ds + \int_{t-\delta}^t \|f(s)\|\,ds \right). \end{aligned}$$

Thus, by condition (1.39), it follows that

(1.40) The function $y_\delta(t)$ converges to the function $y(t)$ as $\delta \downarrow 0$, uniformly in $t \in [0, T]$.

But we have for $0 \leq t < t + h \leq T$ (setting $f(s) = 0$ for $s < 0$)

$$\begin{aligned} y_\delta(t+h) - y_\delta(t) &= \int_0^{t+h-\delta} U(t+h-s)f(s)\,ds - \int_0^{t-\delta} U(t-s)f(s)\,ds \\ &= \int_{t-\delta}^{t+h-\delta} U(t+h-s)f(s)\,ds \\ &\quad + \int_0^{t-\delta} [U(t+h-s) - U(t-s)]f(s)\,ds \end{aligned}$$

$$= [U(h) - I] \int_0^{t-\delta} U(t-s)f(s)\,ds$$
$$+ \int_{t-\delta}^{t+h-\delta} U(t+h-s)f(s)\,ds,$$

so that as $h \downarrow 0$

$$\begin{aligned} &\|y_\delta(t+h) - y_\delta(t)\| \\ &\leq \left\| [U(h) - I] \int_0^{t-\delta} U(t-s)f(s)\,ds \right\| + M_0 \int_{t-\delta}^{t-\delta+h} \|f(s)\|\,ds \\ &\longrightarrow 0. \end{aligned}$$

This proves that

$$y_\delta(t) \in C([0,T]; E).$$

Hence, in view of assertion (1.40), it follows that

$$y(t) \in C([0,T]; E).$$

(b) Next we show that

$$\text{(1.41)} \qquad \begin{cases} y(t) \in D(A), \quad 0 < t < T, \\ Ay(t) = \int_0^t AU(t-s)(f(s) - f(t))\,ds + (U(t) - I)f(t). \end{cases}$$

Since we have by estimate (1.22) and condition (1.39)

$$\begin{aligned} \int_0^{t-\delta} \|AU(t-s)f(s)\|\,ds &\leq M_1 \int_0^{t-\delta} \frac{1}{t-s}\|f(s)\|\,ds \\ &\leq \frac{M_1}{\delta} \int_0^T \|f(s)\|\,ds \\ &< \infty, \end{aligned}$$

in view of the closedness of A it follows that

$$\begin{cases} y_\delta(t) = \int_0^{t-\delta} U(t-s)f(s) \in D(A) \quad \text{for } \delta < t < T, \\ Ay_\delta(t) = \int_0^{t-\delta} AU(t-s)f(s)\,ds. \end{cases}$$

We remark that

$$\begin{aligned} Ay_\delta(t) &= \int_0^{t-\delta} AU(t-s)(f(s) - f(t))\,ds + \int_0^{t-\delta} AU(t-s)f(t)\,ds \\ &= \int_0^{t-\delta} AU(t-s)(f(s) - f(t))\,ds - \int_0^{t-\delta} \frac{d}{ds}(U(t-s)f(t))\,ds \end{aligned}$$

$$= \int_0^{t-\delta} AU(t-s)(f(s)-f(t))\,ds + (U(t)-U(\delta))f(t).$$

Let $[a,b]$ be an arbitrary closed interval of $(0,T)$, and let L_{ab} be a Hölder constant for the function f on the interval $[a/2,b]$:

$$\|f(t)-f(s)\| \le L_{ab}|t-s|^\gamma, \quad t,s \in [a/2,b].$$

Then, by estimate (1.22), we have for $0 < \delta' < \delta < a/2$

$$\begin{aligned}
&\left\| \int_{t-\delta}^{t-\delta'} AU(t-s)(f(s)-f(t))\,ds \right\| \\
&\le M_1 L_{ab} \int_{t-\delta}^{t-\delta'} (t-s)^{\gamma-1}\,ds \\
&= \frac{M_1 L_{ab}}{\gamma}(\delta^\gamma - \delta'^\gamma), \quad t \in [a,b].
\end{aligned}$$

This proves that

The improper integral

$$(1.42) \quad \int_0^t AU(t-s)(f(s)-f(t))\,ds = \lim_{\delta\downarrow 0} \int_0^{t-\delta} AU(t-s)(f(s)-f(t))\,ds$$

exists, and the convergence is uniform in $t \in [a,b] \subset (0,T)$.

Hence we have, as $\delta \downarrow 0$,

$$(1.43) \quad Ay_\delta(t) \longrightarrow \int_0^t AU(t-s)(f(s)-f(t))\,ds + (U(t)-I)f(t), \quad 0 < t < T.$$

Therefore, assertion (1.41) follows from assertions (1.40) and (1.43).

(c) Finally we show that

$$\begin{cases} y(t) \in C^1((0,T);E), \\ y'(t) = Ay(t) + f(t), \quad 0 < t < T. \end{cases}$$

First we remark that the function $y_\delta(t)$ is continuously differentiable on the interval (δ,T) and satisfies

$$\begin{aligned}
y'_\delta(t) &= \int_0^{t-\delta} \frac{d}{dt}(U(t-s)f(s))\,ds + U(\delta)f(t-\delta) \\
&= \int_0^{t-\delta} AU(t-s)f(s)\,ds + U(\delta)f(t-\delta),
\end{aligned}$$

since the semigroup $U(t)$ is real analytic for $t > 0$. Thus we have for $\delta < t < T$

$$\begin{aligned} y'_\delta(t) &= \int_0^{t-\delta} AU(t-s)(f(s)-f(t))\,ds + U(\delta)f(t-\delta) \\ &\quad + \int_0^{t-\delta} AU(t-s)f(t)\,ds \\ &= \int_0^{t-\delta} AU(t-s)(f(s)-f(t))\,ds + U(\delta)f(t-\delta) \\ &\quad - \int_0^{t-\delta} \frac{d}{ds}(U(t-s)f(t))\,ds \\ &= \int_0^{t-\delta} AU(t-s)(f(s)-f(t))\,ds + U(\delta)(f(t-\delta)-f(t)) + U(t)f(t). \end{aligned}$$

But, by estimate (1.21), we have for any closed interval $[a,b]$ of $(0,T)$

$$\|U(\delta)(f(t-\delta)-f(t))\| \leq M_0 L_{ab}\delta^\gamma, \quad t \in [a,b], \tag{1.44}$$

where $0 < \delta < a/2$ and $L_{ab} > 0$ is a Hölder constant for the function f on the interval $[a/2, b]$.

Therefore, combining assertions (1.42) and (1.44), we find that as $\delta \downarrow 0$

$$y'_\delta(t) \longrightarrow \int_0^t AU(t-s)(f(s)-f(t))\,ds + U(t)f(t) = Ay(t) + f(t),$$

uniformly in t over closed intervals of $(0,T)$.

Thus one can let $\delta \downarrow 0$ in the formula

$$y_\delta(t) = \int_\varepsilon^t y'_\delta(\tau)\,d\tau + y_\delta(\varepsilon), \quad 0 < \delta < \varepsilon,$$

to obtain that

$$y(t) = \int_\varepsilon^t (Ay(\tau) + f(\tau))d\tau + y(\varepsilon), \quad 0 < \varepsilon \leq t < T.$$

Since ε is arbitrary, this proves that

$$y(t) \in C^1((0,T);E),$$

and also

$$y'(t) = Ay(t) + f(t), \quad 0 < t < T.$$

Summing up, we have proved that the function $y(t)$, defined by formula (1.36′), belongs to $C([0,T];E) \cap C^1((0,T);E)$, and is a (unique) solution of problem (NP) with initial condition $y(0) = 0$.

The proof of Theorem 1.16 is now complete. □

1.4 The Semilinear Cauchy Problem

Assume that the operator A satisfies condition (1.17). Then we can define the fractional power $(-A)^\alpha$ for $0 < \alpha < 1$. The operator $(-A)^\alpha$ is a closed linear, invertible operator with domain $D((-A)^\alpha) \supset D(A)$.

We let

E_α = the space $D((-A)^\alpha)$ endowed with the graph norm $\|\cdot\|_\alpha$ of $(-A)^\alpha$,

where

$$\|x\|_\alpha = \left(\|x\|^2 + \|(-A)^\alpha x\|^2\right)^{1/2}, \quad x \in D((-A)^\alpha).$$

Then we have the following:

Proposition 1.17. *(i) The space E_α is a Banach space.*
(ii) The graph norm $\|x\|_\alpha$ is equivalent to the norm $\|(-A)^\alpha x\|$.
(iii) If $0 < \alpha < \beta < 1$, then we have $E_\beta \subset E_\alpha$ with continuous injection.

Proof. (i) Assume that $\{x_j\}$ is a Cauchy sequence in E_α, that is,

$$\begin{cases} \{x_j\} \text{ is a Cauchy sequence in } E, \\ \{(-A)^\alpha x_j\} \text{ is a Cauchy sequence in } E. \end{cases}$$

Then there exist elements $x, y \in E$ such that

$$\begin{cases} x_j \to x \text{ in } E, \\ (-A)^\alpha x_j \to y \text{ in } E. \end{cases}$$

Hence, by the closedness of $(-A)^\alpha$, we have

$$\begin{cases} x \in D((-A)^\alpha), \\ (-A)^\alpha x = y. \end{cases}$$

This proves that

$$\begin{cases} x \in E_\alpha, \\ x_j \to x \text{ in } E_\alpha. \end{cases}$$

(ii) Recall that

$$\|(-A)^{-\alpha}\| \leq M, \quad 0 < \alpha < 1. \tag{1.20}$$

Hence we have for all $x \in ((-A)^\alpha)$

$$\|(-A)^\alpha x\|^2 \leq \|x\|^2 + \|(-A)^\alpha x\|^2 \leq (M^2 + 1)\|(-A)^\alpha x\|^2,$$

so that

$$\|(-A)^\alpha x\| \leq \|x\|_\alpha \leq (M^2 + 1)^{1/2}\|(-A)^\alpha x\|, \quad x \in E_\alpha. \tag{1.45}$$

This proves part (ii).

(iii) We remark that

$$(-A)^{\alpha} = (-A)^{-(\beta-\alpha)}(-A)^{\beta}.$$

Hence we have $D((-A)^{\beta}) \subset D((-A)^{\alpha})$. Furthermore, in view of inequalities (1.20) and (1.45), it follows that, for all $x \in E_\beta = D((-A)^{\beta})$,

$$\begin{aligned} \|x\|_\alpha &\leq (1+M^2)^{1/2}\|(-A)^{\alpha}x\| \\ &\leq M(1+M^2)^{1/2}\|(-A)^{\beta}x\| \\ &\leq M(1+M^2)^{1/2}\|x\|_\beta. \end{aligned}$$

This proves part (iii).

The proof of Proposition 1.17 is complete. □

Now we consider the following *semilinear* Cauchy problem:

$$\text{(SLP)} \qquad \begin{cases} \frac{du}{dt} = Au(t) + f(t, u(t)), & t_0 < t < t_1, \\ u(t_0) = x_0. \end{cases}$$

Here f is a function defined on an open subset U of $[0, \infty) \times E_\alpha$, $0 < \alpha < 1$, taking values in E. We assume that $f(t, x)$ is locally Hölder continuous in t and locally Lipschitz continuous in x. That is, for each point (t, x) of U, there exist a neighborhood $V \subset U$, constants $L = L(t, x, V) > 0$ and $0 < \gamma \leq 1$ such that

$$\text{(1.46)} \qquad \|f(s_1, y_1) - f(s_2, y_2)\| \leq L\left(|s_1 - s_2|^{\gamma} + \|y_1 - y_2\|_\alpha\right), \quad (s_1, y_1), (s_2, y_2) \in V.$$

A function $u(t) : [t_0, t_1) \longrightarrow E$ is called a *solution* of problem (SLP) if it satisfies the following three conditions:

(1) $u(t) \in C([t_0, t_1); E) \cap C^1((t_0, t_1); E)$ and $u(t_0) = x_0$.
(2) $u(t) \in D(A)$ and $(t, u(t)) \in U$ for all $t_0 < t < t_1$.
(3) $\frac{du}{dt} = Au(t) + f(t, u(t))$ for all $t_0 < t < t_1$.

Our main result is the following local existence and uniqueness theorem (cf. [He, Theorem 3.3.3]; [Pa, Chapter 6, Theorem 3.1]):

Theorem 1.18. *Let f be a function defined on an open subset U of $[0, \infty) \times E_\alpha$, $0 < \alpha < 1$, taking values in E. Assume that $f(t, x)$ is locally Hölder continuous in t and locally Lipschitz continuous in x. Then, for every $(t_0, x_0) \in U$, problem (CP) has a unique local solution $u(t)$ in the space $C([t_0, t_1]; E) \cap C^1((t_0, t_1); E)$ where $t_1 = t_1(t_0, x_0) > t_0$.*

Proof. (1) We fix a point (t_0, x_0) of U, and choose constants $\varepsilon > 0$ and $\delta > 0$ such that

$$V = \{(t, x) \in [0, \infty) \times E_\alpha : t_0 \leq t \leq t_0 + \varepsilon, \|(-A)^{\alpha}x - (-A)^{\alpha}x_0\| \leq \delta\} \subset U,$$

and

$$\|f(t,x) - f(s,y)\| \leq L\left(|t-s|^\gamma + \|(-A)^\alpha x - (-A)^\alpha y\|\right), \quad (t,x),(s,y) \in V. \tag{1.46'}$$

Here $L = L(t_0, x_0, V) > 0$ is a local Hölder constant for the function f. By part (c) of Theorem 1.12, it follows that, for some constant $M_\alpha > 0$,

$$\|(-A)^\alpha U(t)\| \leq M_\alpha t^{-\alpha}, \quad t > 0. \tag{1.47}$$

Now choose a real number t_1 such that

$$0 < t_1 - t_0 < \min\left\{\varepsilon, \left(\frac{\delta}{2}(1-\alpha)\frac{1}{M_\alpha(B+\delta L)}\right)^{1/(1-\alpha)}\right\}, \tag{1.48}$$

where

$$B = \max_{t_0 \leq t \leq t_0 + \varepsilon} \|f(t, x_0)\|. \tag{1.49}$$

Further, one may assume that $t_1 - t_0$ is so small that

$$\|U(t-t_0)(-A)^\alpha x_0 - (-A)^\alpha x_0\| < \frac{\delta}{2}, \quad t_0 \leq t < t_1. \tag{1.50}$$

(2) We let

Y = the space $C([t_0, t_1]; E)$ of continuous functions on the interval $[t_0, t_1]$ taking values in E.

The space Y is a Banach space with the maximum norm

$$|||y||| = \max_{t_0 \leq t \leq t_1} \|y(t)\|.$$

We define a mapping

$$\Phi : Y \longrightarrow Y$$

as follows:

$$\Phi(y) = U(t-t_0)(-A)^\alpha x_0 + \int_{t_0}^t (-A)^\alpha U(t-s) f\left(s, (-A)^{-\alpha} y(s)\right) ds.$$

We have to verify the well-definedness of the mapping Φ.

(a) First, by Theorem 1.16, it follows that

$$U(t-t_0)(-A)^\alpha x_0 \in C([t_0, \infty); E) \bigcap C^\infty((t_0, \infty); E).$$

(b) Next we show that, for $0 < \beta < 1 - \alpha$,

$$\int_{t_0}^{t} (-A)^\alpha U(t-s) f(s, (-A)^{-\alpha} y(s))\, ds \in C^\beta([t_0, t_1]; E), \tag{1.51}$$

which proves that

$$\Phi y(t) \in C([t_0, t_1]; E) \bigcap C^\beta((t_0, t_1]; E), \quad 0 < \beta < 1 - \alpha. \tag{1.52}$$

We have for $t_0 \le t < t + h \le t_1$

$$\begin{aligned} &\int_{t_0}^{t+h} (-A)^\alpha U(t+h-s) f(s, (-A)^{-\alpha} y(s))\, ds \\ &\quad - \int_{t_0}^{t} (-A)^\alpha U(t-s) f(s, (-A)^{-\alpha} y(s))\, ds \\ &= \int_{t_0}^{t} (-A)^\alpha [U(t+h-s) - U(t-s)] f(s, (-A)^{-\alpha} y(s))\, ds \\ &\quad + \int_{t}^{t+h} (-A)^\alpha U(t+h-s) f(s, (-A)^{-\alpha} y(s))\, ds. \end{aligned}$$

We estimate each term on the right of the last equality.

(b-1) By the continuity of $y(t)$ and inequality (1.46′), it follows that $f(t, (-A)^{-\alpha} y(t))$ is continuous on $[t_0, t_1]$. Hence there exists a constant $N > 0$ such that

$$\|f(t, (-A)^{-\alpha} y(t))\| \le N, \quad t \in [t_0, t_1]. \tag{1.53}$$

Further, by part (d) of Theorem 1.12, we have for each $0 < \beta < 1 - \alpha$

$$\|(U(h) - I)(-A)^\alpha U(t-s)\| \le C_\beta h^\beta \|(-A)^{\alpha+\beta} U(t-s)\|, \quad h > 0, \tag{1.54}$$

with a constant $C_\beta > 0$. Thus, combining inequalities (1.54) and (1.47), we obtain that

$$\|(U(h) - I)(-A)^\alpha U(t-s)\| \le C_\beta h^\beta M_{\alpha+\beta} (t-s)^{-\alpha-\beta}, \quad h > 0. \tag{1.55}$$

Hence, it follows from inequalities (1.55) and (1.53) that

$$\begin{aligned} &\left\| \int_{t_0}^{t} (-A)^\alpha [U(t+h-s) - U(t-s)] f\left(s, (-A)^{-\alpha} y(s)\right) ds \right\| \\ &\le C_\beta M_{\alpha+\beta} N \int_{t_0}^{t} (t-s)^{-\alpha-\beta}\, ds \cdot h^\beta \\ &\le \left(\frac{C_\beta M_{\alpha+\beta} N}{1 - \alpha - \beta} (t_1 - t_0)^{1-\alpha-\beta} \right) h^\beta, \quad 0 < \beta < 1 - \alpha. \end{aligned} \tag{1.56}$$

(b-2) Similarly, we have

$$\left\| \int_t^{t+h} (-A)^\alpha U(t+h-s) f\left(s, (-A)^{-\alpha} y(s)\right) ds \right\| \tag{1.57}$$
$$\leq N \int_t^{t+h} \|(-A)^\alpha U(t+h-s)\| \, ds$$
$$\leq N M_\alpha \int_t^{t+h} (t+h-s)^{-\alpha} \, ds$$
$$= \left(\frac{N M_\alpha}{1-\alpha} \right) h^{1-\alpha}.$$

Therefore, assertion (1.51) follows immediately from estimates (1.56) and (1.57).

(3) We let

$$S = \left\{ y \in Y : y(t_0) = (-A)^\alpha x_0, \max_{t_0 \leq t \leq t_1} \|y(t) - (-A)^\alpha x_0\| \leq \delta \right\}.$$

We remark that S is a non-empty closed and bounded subset of Y, and so it is a complete metric space.

We show the following:

Claim I. $\Phi : S \to S$.

To show this, it suffices to verify that

$$\|\Phi y(t) - (-A)^\alpha x_0\| \leq \delta, \quad t_0 \leq t \leq t_1. \tag{1.58}$$

By (1.50), (1.47), (1.46′) and (1.49), we have for all $t \in [t_0, t_1]$

$$\|\Phi y(t) - (-A)^\alpha x_0\|$$
$$= \left\| U(t-t_0)(-A)^\alpha x_0 - (-A)^\alpha x_0 \right.$$
$$\left. + \int_{t_0}^t (-A)^\alpha U(t-s) f\left(s, (-A)^{-\alpha} y(s)\right) ds \right\|$$
$$\leq \|U(t-t_0)(-A)^\alpha x_0 - (-A)^\alpha x_0\|$$
$$+ \left\| \int_{t_0}^t (-A)^\alpha U(t-s) \left[f\left(s, (-A)^{-\alpha} y(s)\right) - f(s, x_0) \right] ds \right\|$$
$$+ \left\| \int_{t_0}^t (-A)^\alpha U(t-s) f(s, x_0) \, ds \right\|$$
$$\leq \frac{\delta}{2} + M_\alpha L \int_{t_0}^t (t-s)^{-\alpha} \|y(s) - (-A)^\alpha x_0\| \, ds$$

$$+M_\alpha B \int_{t_0}^{t} (t-s)^{-\alpha}\, ds$$
$$\leq \frac{\delta}{2} + M_\alpha (L\delta + B) \int_{t_0}^{t} (t-s)^{-\alpha}\, ds$$
$$\leq \frac{\delta}{2} + \frac{M_\alpha (L\delta + B)}{1-\alpha} (t_1 - t_0)^{1-\alpha}.$$

In view of condition (1.48), this proves inequality (1.58) and hence Claim I. □

Next we show the following:

Claim II. $|||\Phi y_1 - \Phi y_2||| \leq \frac{1}{2}|||y_1 - y_2|||$, $y_1, y_2 \in S$.

By (1.47), (1.46′) and (1.48), we have for all $t \in [t_0, t_1]$

$$\|\Phi y_1(t) - \Phi y_2(t)\|$$
$$= \left\| \int_{t_0}^{t} (-A)^\alpha U(t-s) \left[f\left(s, (-A)^{-\alpha} y_1(s)\right) - f\left(s, (-A)^{\alpha} y_2(s)\right) \right] ds \right\|$$
$$\leq \int_{t_0}^{t} \|(-A)^\alpha U(t-s)\| \cdot \left\| f\left(s, (-A)^{-\alpha} y_1(s)\right) - f\left(s, (-A)^{\alpha} y_2(s)\right) \right\| ds$$
$$\leq M_\alpha L \int_{t_0}^{t} (t-s)^{-\alpha} \|y_1(s) - y_2(s)\|\, ds$$
$$\leq M_\alpha L \frac{(t_1 - t_0)^{1-\alpha}}{1-\alpha} |||y_1 - y_2|||$$
$$\leq \frac{1}{2} |||y_1 - y_2|||. \quad \square$$

Therefore, by Claims I and II, one can apply the contraction mapping theorem to Φ to obtain that

There exists a unique fixed point $y \in S$ of the mapping Φ, that is, $\Phi y = y$, or equivalently,

$$(1.59) \quad y(t) = U(t-t_0)(-A)^\alpha x_0 + \int_{t_0}^{t} (-A)^\alpha U(t-s) f(s, (-A)^{-\alpha} y(s))\, ds.$$

Since $y = \Phi y$, it follows from assertion (1.52) that

For each $t_0' > t_0$, there exists a constant $L_{t_0'} > 0$ such that

$$\|y(t) - y(s)\| \leq L_{t_0'} |t-s|^\beta, \quad t, s \in [t_0', t_1].$$

By inequality (1.46′), this implies the local Hölder continuity of the function $f(t, (-A)^{-\alpha} y(t))$. Indeed, we have for all] $t, s \in [t_0', t_1]$, $t_0 < t_0' < t_1$,

$$\left\| f\left(t, (-A)^{-\alpha} y(t)\right) - f\left(s, (-A)^{-\alpha} y(s)\right) \right\|$$

$$\leq L\left(|t-s|^\gamma + \|y(t)-y(s)\|\right)$$
$$\leq L|t-s|^\gamma + LL_{t_0'}|t-s|^\beta, \quad 0<\gamma\leq 1,\ 0<\beta<1-\alpha.$$

(4) Now we consider the following non-homogeneous Cauchy problem:

$$\text{(NP)} \qquad \begin{cases} \frac{dx}{dt} = Ax(t) + f\left(t, (-A)^{-\alpha}y(t)\right), & 0<t<T, \\ x(0) = x_0. \end{cases}$$

But we know that $f(t,(-A)^{-\alpha}y(t))$ belongs to $C([t_0,t_1];E) \cap C^{\beta'}((t_0,t_1];E)$ where $0<\beta'<\min(\gamma, 1-\alpha)$. Hence, applying Theorem 1.16, we obtain that the function

$$u(t) = U(t-t_0)x_0 + \int_{t_0}^t U(t-s)f(s,(-A)^{-\alpha}y(s))\,ds$$

belongs to $C([t_0,t_1];E)\cap C^1((t_0,t_1);E)$, and is a unique solution of problem (NP).

It remains to show that

$$u(t) = (-A)^{-\alpha}y(t), \quad t_0<t\leq t_1, \tag{1.60}$$

which proves that the function u is a solution of problem (SLP).

Since we have for all $t_0<t\leq t_1$

$$\begin{cases} U(t-t_0)x_0 \in D(A) \subset D\left((-A)^\alpha\right), \\ \int_t^{t_0} U(t-s)f\left(s,(-A)^{-\alpha}y(s)\right)ds \in D(A) \subset D\left((-A)^\alpha\right), \end{cases}$$

in view of part (b) of Theorem 1.12 and formula (1.59) it follows that

$$\begin{aligned}(-A)^\alpha u(t) &= (-A)^\alpha U(t-t_0)x_0 + \int_{t_0}^t (-A)^\alpha U(t-s)f(s,(-A)^{-\alpha}y(s))\,ds \\ &= U(t-t_0)(-A)^\alpha x_0 + \int_{t_0}^t (-A)^\alpha U(t-s)f(s,(-A)^{-\alpha}y(s))\,ds \\ &= y(t).\end{aligned}$$

This proves formula (1.60).

(5) Finally, we prove the uniqueness of solutions of problem (SLP). Assume that u, v are two solutions of problem (SLP):

$$\begin{cases} \frac{du}{dt} = Au(t) + f(t,u(t)), \ t_0<t<t_1, \\ u(t_0) = x_0; \end{cases}$$

$$\begin{cases} \frac{dv}{dt} = Av(t) + f(t,v(t)), \ t_0<t<t_1, \\ v(t_0) = x_0. \end{cases}$$

Then we have by Theorem 1.15

$$u(t) = U(t - t_0)x_0 + \int_{t_0}^{t} U(t - s)f(s, u(s))\, ds,$$
$$v(t) = U(t - t_0)x_0 + \int_{t_0}^{t} U(t - s)f(s, v(s))\, ds,$$

and hence

$$(-A)^\alpha u(t) = U(t - t_0)(-A)^\alpha x_0 + \int_{t_0}^{t} (-A)^\alpha U(t - s)f(s, u(s))\, ds,$$
$$(-A)^\alpha v(t) = U(t - t_0)(-A)^\alpha x_0 + \int_{t_0}^{t} (-A)^\alpha U(t - s)f(s, v(s))\, ds.$$

This implies that the functions $(-A)^\alpha u(t)$ and $(-A)^\alpha v(t)$ are both fixed points of the mapping Φ. (Here we remark that the interval $[t_0, t_1]$ should be replaced by a small interval $[t_0, t_1 - h]$, $h > 0$.) Thus, by the uniqueness of fixed points of Φ, it follows that

$$(-A)^\alpha u(t) = (-A)^\alpha v(t), \quad t_0 \leq t < t_1,$$

so that

$$u(t) = v(t), \quad t_0 \leq t < t_1.$$

The proof of Theorem 1.18 is now complete. □

CHAPTER II

SOBOLEV IMBEDDING THEOREMS

In this chapter we study the imbedding characteristics of Sobolev spaces of L^p style that render these spaces so useful in the study of partial differential equations. We give a complete proof of the most important of the imbedding properties of the spaces $H^{m,p}(\mathbf{R}^n)$ (Theorems 2.15 and 2.18 and Corollary 2.22). For more leisurely treatments of Sobolev spaces, the reader might refer to Adams [Ad] and also Friedman [Fr].

2.1 Hölder Spaces and Sobolev Spaces

Let Ω be an open subset of $\mathbf{R}^n$. If m is a non-negative integer, we let

$C^m(\Omega) =$ the space of m times continuously differentiable functions u in Ω, equipped with the topology of uniform convergence on every compact subset of Ω of the functions u and their derivatives $D^\alpha u$ of order $|\alpha| \leq m$,

and

$B^m(\Omega) =$ the subspace of $C^m(\Omega)$ consisting of functions u having all their derivatives $D^\alpha u$, $|\alpha| \leq m$, bounded in the whole of Ω, equipped with the topology of uniform convergence over Ω of the functions u and their derivatives $D^\alpha u$.

On the space $B^m(\Omega)$, we introduce seminorms

$$|u|_{j,\infty,\Omega} = \max_{|\alpha|=j} \sup_{x\in\Omega} |D^\alpha u(x)|, \quad 0 \leq j \leq m,$$

and a norm

$$\|u\|_{m,\infty,\Omega} = \max_{0\leq j\leq m} |u|_{j,\infty,\Omega}.$$

We remark that

$$\|u\|_{0,\infty,\Omega} = |u|_{0,\infty,\Omega} = \sup_{x\in\Omega} |u(x)|,$$

Typeset by $\mathcal{A}_\mathcal{M}\mathcal{S}$-TEX

$$|u|_{m,\infty,\Omega} = \max_{|\alpha|=m} |D^\alpha u|_{0,\infty,\Omega}.$$

If $\Omega = \mathbf{R}^n$, we simply write

$$|\cdot|_{m,\infty} = |\cdot|_{m,\infty,\mathbf{R}^n},$$
$$\|\cdot\|_{m,\infty} = \|\cdot\|_{m,\infty,\mathbf{R}^n}.$$

Let $0 < \theta < 1$. A function u defined on Ω is said to be *uniformly Hölder continuous* with exponent θ on Ω if the quantity

$$[u]_{\theta;\Omega} = \sup_{\substack{x,y\in\Omega \\ x\neq y}} \frac{|u(x)-u(y)|}{|x-y|^\theta}$$

is finite.

If $a = m + \theta$, where m is a non-negative integer and $0 < \theta < 1$, we let

$B^a(\Omega) =$ the space of functions in $B^m(\Omega)$ whose m-th order derivatives are uniformly Hölder continuous with exponent θ on Ω.

We define a seminorm

$$|u|_{a,\infty,\Omega} = \max_{|\alpha|=m} [D^\alpha u]_{\theta;\Omega} = \max_{|\alpha|=m} \sup_{\substack{x,y\in\Omega \\ x\neq y}} \frac{|D^\alpha u(x) - D^\alpha u(y)|}{|x-y|^\theta},$$

and a norm

$$\|u\|_{a,\infty,\Omega} = \max\left\{\|u\|_{m,\infty,\Omega}, |u|_{a,\infty,\Omega}\right\}.$$

If $\Omega = \mathbf{R}^n$, we simply write

$$|\cdot|_{a,\infty} = |\cdot|_{a,\infty,\mathbf{R}^n},$$
$$\|\cdot\|_{a,\infty} = \|\cdot\|_{a,\infty,\mathbf{R}^n}.$$

If m is a non-negative integer, we define the Sobolev space

$H^{m,p}(\Omega) =$ the space of (equivalence classes of) functions $u \in L^p(\Omega)$ whose distribution derivatives $D^\alpha u$ of order $|\alpha| \leq m$ are all in $L^p(\Omega)$.

We introduce seminorms $|\cdot|_{k,p,\Omega}$, $0 \leq k \leq m$, as follows:

$$|u|_{k,p,\Omega} = \begin{cases} \left(\int_\Omega \sum_{|\alpha|=k} |D^\alpha u(x)|^p \, dx\right)^{1/p} & \text{if } 1 \leq p < \infty, \\ \max_{|\alpha|=k} \operatorname{ess\,sup}_{x\in\Omega} |D^\alpha u(x)| & \text{if } p = \infty. \end{cases}$$

If $\Omega = \mathbf{R}^n$, we simply write

$$|\cdot|_{k,p} = |\cdot|_{k,p,\mathbf{R}^n}.$$

Furthermore, we introduce a norm

$$\|u\|_{m,p,\Omega} = \begin{cases} \left(\int_\Omega \sum_{|\alpha|\le m} |D^\alpha u(x)|^p \, dx\right)^{1/p} & \text{if } 1 \le p < \infty, \\ \max_{|\alpha|\le m} \text{ess sup}_{x\in\Omega} |D^\alpha u(x)| & \text{if } p = \infty \end{cases}$$
$$= \begin{cases} \left(\sum_{k=0}^m |u|_{k,p,\Omega}^p\right)^{1/p} & \text{if } 1 \le p < \infty, \\ \max_{0\le k\le m} |u(x)|_{k,\infty,\Omega} & \text{if } p = \infty. \end{cases}$$

We remark that

$$H^{0,p}(\Omega) = L^p(\Omega), \quad \|\cdot\|_{0,p,\Omega} = |\cdot|_{0,p,\Omega}.$$

If $\Omega = \mathbf{R}^n$, we simply write

$$\|\cdot\|_{m,p} = \|\cdot\|_{m,p,\mathbf{R}^n}.$$

2.2 Interpolation Theorems

In this section, we consider the problem of determining upper bounds for L^r-norms of derivatives $D^\beta u$, $|\beta| < m$, of functions $u \in H^{m,p}(\mathbf{R}^n)$, in terms of the L^q-norms of u and its derivatives $D^\alpha u$ of order $|\alpha| = m$.

Following Friedman [Fr], we define

$$|u|_{p,\Omega} = \begin{cases} \left(\int_\Omega |u(x)|^p \, dx\right)^{1/p} & \text{if } 0 < p < \infty, \\ |u|_{-n/p,\infty,\Omega} & \text{if } -\infty < p < 0, \\ |u|_{0,\infty,\Omega} & \text{if } p = \pm\infty. \end{cases}$$

If $\Omega = \mathbf{R}^n$, we simply write

$$|\cdot|_p = |\cdot|_{p,\mathbf{R}^n}.$$

The purpose of this section is to prove the following:

Theorem 2.1. *Let $1 \le p \le \infty$, $1 \le q \le \infty$ and let j, m be integers such that $0 \le j < m$. If $m - j - n/p$ is not a non-negative integer and if*

$$\frac{1}{r} = \frac{j}{n} + a\left(\frac{1}{p} - \frac{m}{n}\right) + (1-a)\frac{1}{q}, \quad \frac{j}{m} \le a \le 1,$$

then we have, for all functions $u \in C_0^m(\mathbf{R}^n)$,

$$|D^j u|_r \le C |u|_{m,p}^a |u|_{0,q}^{1-a}. \tag{2.1}$$

Here $C = C(m, n, p, q, j, a)$ is a positive constant.

Remark 2.2. Since $a \geq j/m$, it follows that

$$\frac{1}{r} = \frac{m}{n}\left(\frac{j}{m} - a\right) + a\frac{1}{p} + (1-a)\frac{1}{q} \leq a\frac{1}{p} + (1-a)\frac{1}{q} \leq 1.$$

Hence we have $r \geq 1$ or $r < 0$.

The proof of Theorem 2.1 is carried out in a chain of auxiliary lemmas. The proof given here is essentially due to Gagliardo [Ga]. Although it is rather lengthy, the techniques involved are quite elementary and based on little more than simple calculus combined with astute applications of Hölder's inequality.

Lemma 2.3. *Let $0 \leq a < b < c < \infty$. Then there exists a constant $\gamma = \gamma(a,b,c) > 0$ such that*

$$|u|_{b,\infty} \leq \gamma \left(|u|_{c,\infty}\right)^{(b-a)/(c-a)} \left(|u|_{a,\infty}\right)^{(c-b)/(c-a)}, \quad u \in B^c(\mathbf{R}^n). \tag{2.2}$$

Proof. (i) The case $0 \leq a < b < c \leq 1$: We remark that

$$\frac{|u(x)-u(y)|}{|x-y|^b} = \left(\frac{|u(x)-u(y)|}{|x-y|^c}\right)^{(b-a)/(c-a)} \left(\frac{|u(x)-u(y)|}{|x-y|^a}\right)^{(c-b)/(c-a)} \tag{2.3}$$

Hence, if $0 < a < b < c < 1$, then we have

$$|u|_{b,\infty} \leq \left(|u|_{c,\infty}\right)^{(b-a)/(c-a)} \left(|u|_{a,\infty}\right)^{(c-b)/(c-a)}.$$

Thus we have only to consider the following two cases:

$$a = 0 \quad \text{and} \quad c = 1.$$

(i-1) The case $c = 1$: We have for $u \in B^1(\mathbf{R}^n)$

$$\begin{aligned} |u(x) - u(y)| &= \left| \int_0^1 \frac{d}{dt}(u(y + t(x-y)))\,dt \right| \\ &= \left| \int_0^1 \sum_{i=1}^n \frac{\partial u}{\partial x_i}(y + t(x-y))(x_i - y_i)\,dt \right| \\ &\leq \sum_{i=1}^n |D_i u|_{0,\infty} |x_i - y_i| \\ &\leq |u|_{1,\infty} \sqrt{n} |x-y|. \end{aligned}$$

But we remark that

$$\left|\frac{\partial u}{\partial x_i}(x)\right| = \lim_{h \to 0} \left|\frac{u(x + he_i) - u(x)}{h}\right| \leq \sup_{x \neq y} \frac{|u(x) - u(y)|}{|x-y|},$$

so that

$$|u|_{1,\infty} = \max_{1\le i\le n} \sup_{x\in\mathbf{R}^n} \left|\frac{\partial u}{\partial x_i}(x)\right| \le \sup_{x\ne y} \frac{|u(x)-u(y)|}{|x-y|}.$$

Summing up, we have proved that

$$|u|_{1,\infty} \le \sup_{x\ne y} \frac{|u(x)-u(y)|}{|x-y|} \le \sqrt{n}|u|_{1,\infty}, \quad u \in B^1(\mathbf{R}^n). \tag{2.4}$$

In view of formula (2.3) with $c = 1$, this proves that

$$|u|_{b,\infty} \le \left(\sqrt{n}|u|_{1,\infty}\right)^{(b-a)/(1-a)} \left(|u|_{a,\infty}\right)^{(1-b)/(1-a)}, \quad u \in B^1(\mathbf{R}^n).$$

(i-2) The case $a = 0$: Since we have

$$|u(x)-u(y)| \le 2|u|_{0,\infty},$$

it follows that

$$|u|_{b,\infty} \le \left(|u|_{c,\infty}\right)^{b/c} \left(2|u|_{0,\infty}\right)^{(c-b)/c}, \quad u \in B^c(\mathbf{R}^n).$$

(ii) The case $0 \le a < b = 1 < c \le 2$: We have for all $\rho > 0$

$$\frac{\partial u}{\partial x_i}(x) = \frac{1}{\rho}\int_{x_i}^{x_i+\rho} \left(\frac{\partial u}{\partial x_i}(x) - \frac{\partial u}{\partial y_i}(y)\right) dy_i + \frac{1}{\rho}\left(u(x+\rho e_i) - u(x)\right).$$

But it follows that

$$|u(x+\rho e_i) - u(x)| \le \begin{cases} |u|_{a,\infty}\rho^a & \text{if } a > 0, \\ 2|u|_{0,\infty} & \text{if } a = 0, \end{cases}$$

and that

$$\left|\frac{\partial u}{\partial x_i}(x) - \frac{\partial u}{\partial y_i}(y)\right| \le \begin{cases} |u|_{c,\infty}|y_i - x_i|^{c-1} & \text{if } 1 < c < 2, \\ \sqrt{n}|u|_{c,\infty}|y_i - x_i|^{c-1} & \text{if } c = 2. \end{cases}$$

Thus, if $a > 0$, we have

$$\begin{aligned} \left|\frac{\partial u}{\partial x_i}(x)\right| &\le \frac{1}{\rho}\int_{x_i}^{x_i+\rho} \sqrt{n}|u|_{c,\infty}|y_i - x_i|^{c-1}\, dy_i + |u|_{a,\infty}\rho^{a-1} \\ &= \frac{\sqrt{n}}{c}|u|_{c,\infty}\rho^{c-1} + |u|_{a,\infty}\rho^{a-1}, \end{aligned}$$

and hence

$$|u|_{1,\infty} \le \frac{\sqrt{n}}{c}|u|_{c,\infty}\rho^{c-1} + |u|_{a,\infty}\rho^{a-1} \quad \text{for all } \rho > 0. \tag{2.5}$$

Therefore, inequality (2.2) for $a > 0$ follows by minimizing the right-hand side of (2.5) with respect to ρ:

$$|u|_{1,\infty} \leq \gamma \left(|u|_{c,\infty}\right)^{(1-a)/(c-a)} \left(|u|_{a,\infty}\right)^{(c-1)/(c-a)}, \quad u \in B^c(\mathbf{R}^n).$$

Similarly, if $a = 0$, we have

$$|u|_{1,\infty} \leq \frac{\sqrt{n}}{c} |u|_{c,\infty} \rho^{c-1} + 2|u|_{0,\infty} \rho^{-1} \quad \text{for all } \rho > 0. \tag{2.6}$$

Hence inequality (2.2) for $a = 0$ follows by minimizing the right-hand side of (2.6) with respect to ρ:

$$|u|_{1,\infty} \leq \gamma \left(|u|_{c,\infty}\right)^{1/c} \left(|u|_{0,\infty}\right)^{(c-1)/c}, \quad u \in B^c(\mathbf{R}^n).$$

The proof of Lemma 2.3 is complete. □

Lemma 2.4. *Let $p \leq -n$, $q > 0$. Then we have*

$$|u|_{0,\infty} \leq \gamma \left(|u|_p\right)^{p/(p-q)} \left(|u|_q\right)^{q/(q-p)}, \quad u \in C_0^\infty(\mathbf{R}^n). \tag{2.7}$$

Proof. (i) The case $p < -n$: We remark that

$$|u(x) - u(y)| \leq |u|_{-n/p,\infty} |x-y|^{-n/p} = |u|_p |x-y|^{-n/p},$$

since $0 < -n/p < 1$. Hence it follows that

$$|u(x)| \leq |u|_p |x-y|^{-n/p} + |u(y)|. \tag{2.8}$$

(i-a) If $q \geq 1$, we integrate both sides of inequality (2.8) over the ball $B_\rho(x) = \{y \in \mathbf{R}^n : |y - x| < \rho\}$ to obtain that

$$\begin{aligned}
|u(x)| V_n \rho^n &\leq |u|_p \left(\int_0^\rho r^{-n/p} r^{n-1}\, dr \right) \omega_n + \int_{B_\rho(x)} |u(y)|\, dy \\
&\leq |u|_p \omega_n \frac{\rho^{n-n/p}}{n-(n/p)} + \left(\int_{B_\rho(x)} |u(y)|^q\, dy \right)^{1/q} \left(\int_{B_\rho(x)} 1\, dy \right)^{1-1/q} \\
&\leq |u|_p (nV_n) \frac{\rho^{n-n/p}}{n-(n/p)} + (V_n \rho^n)^{1-1/q} |u|_q.
\end{aligned}$$

Here V_n is the volume of the unit ball and $\omega_n = nV_n$ is the surface area of the unit ball. Hence we have

$$|u|_{0,\infty} \leq |u|_p \frac{\rho^{-n/p}}{1-(1/p)} + (V_n \rho^n)^{-1/q} |u|_q \quad \text{for all } \rho > 0. \tag{2.9}$$

Therefore, inequality (2.7) for $q \geq 1$ follows by minimizing the right-hand side of (2.9) with respect to ρ.

(i-b) If $0 < q < 1$, we find from inequality (2.8) that

$$|u(x)| \leq |u|_p |x-y|^{-n/p} + (|u|_{0,\infty})^{1-q} |u(y)|^q.$$

Hence, integrating both sides over the ball $B_\rho(x)$, we obtain that

$$|u(x)| V_n \rho^n \leq |u|_p (nV_n) \frac{\rho^{n-n/p}}{n-(n/p)} + (|u|_{0,\infty})^{1-q} |u|_q^q,$$

so that

$$|u|_{0,\infty} \leq |u|_p \frac{\rho^{-n/p}}{1-(1/p)} + (V_n \rho^n)^{-1} (|u|_{0,\infty})^{1-q} |u|_q^q. \tag{2.10}$$

Therefore, inequality (2.7) for $0 < q < 1$ follows by minimizing the right-hand side of (2.10) with respect to ρ.

(ii) The case $p = -n$: By inequality (2.4), it follows that

$$\begin{aligned} |u(x)| &\leq |u(x) - u(y)| + |u(y)| \\ &\leq \sqrt{n} |u|_{1,\infty} |x-y| + |u(y)| \\ &= \sqrt{n} |u|_p |x-y| + |u(y)|. \end{aligned}$$

Therefore, inequality (2.7) can be obtained just as in case (i).

The proof of Lemma 2.4 is complete. □

Lemma 2.5. *Let $r < p \leq -n$, $q > 0$. Then we have*

$$|u|_r \leq \gamma (|u|_p)^{p(r-q)/r(p-q)} (|u|_q)^{q(r-p)/r(q-p)}, \quad u \in C_0^\infty(\mathbf{R}^n). \tag{2.11}$$

Proof. We recall that

$$|u|_{1,\infty} \leq \sup_{x \neq y} \frac{|u(x)-u(y)|}{|x-y|} \leq \sqrt{n} |u|_{1,\infty}. \tag{2.4}$$

Since $0 < p/r < 1$, $0 < -n/r < -n/p \leq 1$, it follows that

$$\begin{aligned} \frac{|u(x)-u(y)|}{|x-y|^{-n/r}} &= \left(\frac{|u(x)-u(y)|}{|x-y|^{-n/p}} \right)^{p/r} |u(x)-u(y)|^{(r-p)/r} \\ &\leq \begin{cases} |u|_p^{p/r} (2|u|_{0,\infty})^{(r-p)/r} & \text{if } p < -n, \\ \sqrt{n} |u|_p^{p/r} (2|u|_{0,\infty})^{(r-p)/r} & \text{if } p = -n, \end{cases} \end{aligned}$$

and hence

$$|u|_r \le \gamma\,(|u|_p)^{p/r}\,(|u|_{0,\infty})^{(r-p)/r}. \tag{2.12}$$

But we have by Lemma 2.4

$$|u|_{0,\infty} \le \gamma\,(|u|_p)^{p/(p-q)}\,(|u|_q)^{q/(q-p)}. \tag{2.7}$$

Therefore, combining inequalities (2.12) and (2.7), we obtain that

$$\begin{aligned} |u|_r &\le \gamma\,(|u|_p)^{p/r}\left((|u|_p)^{p/(p-q)}\,(|u|_q)^{q/(q-p)}\right)^{(r-p)/r} \\ &\le \gamma\,(|u|_p)^{p(r-q)/r(p-q)}\,(|u|_q)^{q(r-p)/r(q-p)}. \end{aligned}$$

This proves Lemma 2.5. □

Lemma 2.6. *Let $p \le -n$, $0 < q < r$. Then we have*

$$|u|_r \le \gamma\,(|u|_p)^{p(r-q)/r(p-q)}\,(|u|_q)^{q(r-p)/r(q-p)}, \quad u \in C_0^\infty(\mathbf{R}^n). \tag{2.13}$$

Proof. By Lemma 2.4, it follows that

$$|u|_{0,\infty} \le \gamma\,(|u|_p)^{p/(p-q)}\,(|u|_q)^{q/(q-p)}. \tag{2.7}$$

Hence we have

$$\begin{aligned} |u|_r &= \left(\int_{\mathbf{R}^n} |u(x)|^r\,dx\right)^{1/r} \\ &= \left(\int_{\mathbf{R}^n} |u(x)|^{r-q}|u(x)|^q\,dx\right)^{1/r} \\ &\le (|u|_{0,\infty})^{(r-q)/r}\left(\int_{\mathbf{R}^n} |u(x)|^q\,dx\right)^{1/r} \\ &\le \left(\gamma\,(|u|_p)^{p/(p-q)}\,(|u|_q)^{q/(q-p)}\right)^{(r-q)/r}(|u|_q)^{q/r} \\ &\le \gamma\,(|u|_p)^{p(r-q)/r(p-q)}\,(|u|_q)^{q(r-p)/r(q-p)}. \end{aligned}$$

The proof of Lemma 2.6 is complete. □

Proposition 2.7. *If $\lambda < \mu < \nu$, then there exists a constant $\gamma = \gamma(\lambda,\mu,\nu) > 0$ such that*

$$|u|_{1/\mu} \le \gamma\,\left(|u|_{1/\lambda}\right)^{(\nu-\mu)/(\nu-\lambda)}\left(|u|_{1/\nu}\right)^{(\mu-\lambda)/(\nu-\lambda)}, \quad u \in C_0^\infty(\mathbf{R}^n). \tag{2.14}$$

Proof. (i) The case $0 \le \lambda < \mu < \nu$: We let

$$p = \frac{\mu(\nu - \lambda)}{\lambda(\nu - \mu)}.$$

Then we have

$$1 < p \le \infty,$$

and

$$p' = \frac{p}{p-1} = \frac{\mu(\nu - \lambda)}{\nu(\mu - \lambda)}.$$

We remark that

$$\begin{aligned} \lambda > 0 &\implies p < \infty, \\ \lambda = 0 &\implies p = \infty. \end{aligned}$$

If $\lambda > 0$, by Hölder's inequality, it follows that

$$\begin{aligned} |u|_{1/\mu} &= \left(\int_{\mathbf{R}^n} |u(x)|^{1/\mu}\, dx \right)^{\mu} \\ &= \left(\int_{\mathbf{R}^n} |u(x)|^{(\nu-\mu)/\mu(\nu-\lambda)} |u(x)|^{(\mu-\lambda)/\mu(\nu-\lambda)}\, dx \right)^{\mu} \\ &\le \left(\int_{\mathbf{R}^n} |u(x)|^{1/\lambda}\, dx \right)^{\mu/p} \left(\int_{\mathbf{R}^n} |u(x)|^{1/\nu}\, dx \right)^{\mu/p'} \\ &\le \left(|u|_{1/\lambda} \right)^{(\nu-\mu)/(\nu-\lambda)} \left(|u|_{1/\nu} \right)^{(\mu-\lambda)/(\nu-\lambda)}. \end{aligned}$$

This proves inequality (2.14) for $\gamma = 1$.

If $\lambda = 0$, we have

$$\begin{aligned} |u|_{1/\mu} &= \left(\int_{\mathbf{R}^n} |u(x)|^{(\nu-\mu)/\mu\nu} |u(x)|^{1/\nu}\, dx \right)^{\mu} \\ &\le \left(|u|_{0,\infty} \right)^{(\nu-\mu)/\nu} \left(\int_{\mathbf{R}^n} |u(x)|^{1/\nu}\, dx \right)^{\mu} \\ &= \left(|u|_{1/\lambda} \right)^{(\nu-\mu)/(\nu-\lambda)} \left(|u|_{1/\nu} \right)^{(\mu-\lambda)/(\nu-\lambda)}. \end{aligned}$$

This proves inequality (2.14) for $\gamma = 1$.

(ii) The case $\lambda < \mu < \nu \le 0$: Since we have

$$0 \le -n\nu < -n\mu < -n\lambda,$$

it follows from an application of Lemma 2.3 with $a = -n\nu$, $b = -n\mu$ and $c = -n\lambda$ that

$$|u|_{-n\mu,\infty} \le \gamma \left(|u|_{-n\lambda,\infty} \right)^{(\nu-\mu)/(\nu-\lambda)} \left(|u|_{-n\nu,\infty} \right)^{(\mu-\lambda)/(\nu-\lambda)}.$$

This implies that

$$|u|_{1/\mu} \leq \gamma \left(|u|_{1/\lambda}\right)^{(\nu-\mu)/(\nu-\lambda)} \left(|u|_{1/\nu}\right)^{(\mu-\lambda)/(\nu-\lambda)}.$$

(iii) The case $-1/n \leq \lambda < \mu \leq 0 < \nu$: If $\mu < 0$, we let

$$p = \frac{1}{\lambda}, \quad r = \frac{1}{\mu}, \quad q = \frac{1}{\nu}.$$

Then we have

$$r < p \leq -n, \quad q > 0.$$

Hence it follows from an application of Lemma 2.5 that

$$|u|_{1/\mu} \leq \gamma \left(|u|_{1/\lambda}\right)^{(\nu-\mu)/(\nu-\lambda)} \left(|u|_{1/\mu}\right)^{(\mu-\lambda)/(\nu-\lambda)}.$$

If $\mu = 0$, applying Lemma 2.4, we obtain that

$$\begin{aligned} |u|_{1/\mu} = |u|_{0,\infty} &\leq \gamma \left(|u|_{-n\lambda,\infty}\right)^{\nu/(\nu-\lambda)} \left(|u|_{-n\nu,\infty}\right)^{\lambda/(\lambda-\nu)} \\ &= \gamma \left(|u|_{1/\lambda}\right)^{(\nu-\mu)/(\nu-\lambda)} \left(|u|_{1/\nu}\right)^{(\mu-\lambda)/(\nu-\lambda)}. \end{aligned}$$

(iv) The case $\lambda < -1/n < \mu \leq 0 < \nu$: If we apply part (iii) with $\lambda = -1/n$, we obtain that

$$|u|_{1/\mu} \leq \gamma \left(|u|_{-n}\right)^{n(\nu-\mu)/(n\nu+1)} \left(|u|_{1/\nu}\right)^{(n\mu+1)/(n\nu+1)}.$$

But we have by Lemma 2.3

$$\begin{aligned} |u|_{-n} = |u|_{1,\infty} &\leq \gamma \left(|u|_{-n\lambda,\infty}\right)^{(n\mu+1)/n(\mu-\lambda)} \left(|u|_{-n\mu,\infty}\right)^{(n\lambda+1)/n(\lambda-\mu)} \\ &= \gamma \left(|u|_{1/\lambda}\right)^{(n\mu+1)/n(\mu-\lambda)} \left(|u|_{1/\mu}\right)^{(n\lambda+1)/n(\lambda-\mu)}. \end{aligned}$$

Hence it follows that

$$\begin{aligned} |u|_{1/\mu} &\leq \gamma \left(|u|_{1/\nu}\right)^{(n\mu+1)/(n\nu+1)} \Bigg(\left(|u|_{1/\lambda}\right)^{(n\mu+1)/n(\mu-\lambda)} \\ &\qquad \times \left(|u|_{1/\mu}\right)^{(n\lambda+1)/n(\lambda-\mu)} \Bigg)^{n(\nu-\mu)/(n\nu+1)} \\ &= \gamma \left(|u|_{1/\nu}\right)^{(n\mu+1)/(n\nu+1)} \left(|u|_{1/\lambda}\right)^{\frac{n\mu+1}{n\nu+1}\frac{\nu-\mu}{\mu-\lambda}} \left(|u|_{1/\mu}\right)^{\frac{n\lambda+1}{n\nu+1}\frac{\nu-\mu}{\lambda-\mu}}, \end{aligned}$$

so that

$$|u|_{1/\mu} \leq \gamma \left(|u|_{1/\lambda}\right)^{(\nu-\mu)/(\nu-\lambda)} \left(|u|_{1/\nu}\right)^{(\mu-\lambda)/(\nu-\lambda)}.$$

(v) The case $\lambda < \mu = -1/n < 0 < \nu$: Since $-n\lambda > 1$, applying Lemma 2.3 with $a = 0$, $b = 1$ and $c = -n\lambda$, we obtain that

$$|u|_{-n} = |u|_{1,\infty} \le \gamma (|u|_{-n\lambda,\infty})^{-1/(n\lambda)} (|u|_{0,\infty})^{(-n\lambda-1)/(-n\lambda)}.$$

But we have by Lemma 2.4 with $p = -n$ and $q = 1/\nu$

$$|u|_{0,\infty} \le \gamma (|u|_{-n})^{n\nu/(n\nu+1)} \left(|u|_{1/\nu}\right)^{1/(n\nu+1)}.$$

Hence it follows that

$$|u|_{-n} \le \gamma \left(|u|_{1/\lambda}\right)^{1/(n\lambda)} \left((|u|_{-n})^{n\nu/(n\nu+1)} \left(|u|_{1/\nu}\right)^{1/(n\nu+1)}\right)^{1+1/n\lambda},$$

so that

$$\begin{aligned} |u|_{1/\mu} = |u|_{-n} &\le \gamma \left(|u|_{1/\lambda}\right)^{(n\nu+1)/n(\nu-\lambda)} \left(|u|_{1/\nu}\right)^{-(n\lambda+1)/n(\nu-\lambda)} \\ &= \gamma \left(|u|_{1/\lambda}\right)^{(\nu-\mu)/(\nu-\lambda)} \left(|u|_{1/\nu}\right)^{(\mu-\lambda)/(\nu-\lambda)}. \end{aligned}$$

(vi) The case $\lambda < \mu < -1/n < 0 < \nu$: If we apply part (v) with $\lambda = \mu$, we obtain that

$$|u|_{-n} \le \gamma \left(|u|_{1/\mu}\right)^{(n\nu+1)/n(\nu-\mu)} \left(|u|_{1/\nu}\right)^{-(n\mu+1)/n(\nu-\mu)}.$$

But, since $1 < -n\mu < -n\lambda$, it follows from an application of Lemma 2.3 with $a = 1$, $b = -n\mu$ and $c = -n\lambda$ that

$$\begin{aligned} |u|_{1/\mu} = |u|_{-n\mu,\infty} &\le \gamma (|u|_{-n\lambda,\infty})^{(n\mu+1)/(n\lambda+1)} \left(|u|_{1,\infty}\right)^{-n(\mu-\lambda)/(n\lambda+1)} \\ &= \gamma \left(|u|_{1/\lambda}\right)^{(n\mu+1)/(n\lambda+1)} \left(|u|_{-n}\right)^{-n(\mu-\lambda)/(n\lambda+1)}. \end{aligned}$$

Hence we have

$$\begin{aligned} |u|_{1/\mu} \le \gamma \left(|u|_{1/\lambda}\right)^{(n\mu+1)/(n\lambda+1)} &\Big(\left(|u|_{1/\mu}\right)^{(n\nu+1)/n(\nu-\mu)} \\ &\times \left(|u|_{1/\nu}\right)^{-(n\mu+1)/n(\nu-\mu)}\Big)^{-n(\mu-\lambda)/(n\lambda+1)}, \end{aligned}$$

so that

$$|u|_{1/\mu} \le \gamma \left(|u|_{1/\lambda}\right)^{(\nu-\mu)/(\nu-\lambda)} \left(|u|_{1/\nu}\right)^{(\mu-\lambda)/(\nu-\lambda)}.$$

(vii) The case $-1/n \le \lambda < 0 < \mu < \nu$: Since we have

$$0 < \frac{1}{\nu} < \frac{1}{\mu}, \quad \frac{1}{\lambda} \le -n,$$

it follows from an application of Lemma 2.6 with $p = 1/\lambda$, $q = 1/\nu$ and $r = 1/\mu$ that

$$|u|_{1/\mu} \leq \gamma \left(|u|_{1/\lambda}\right)^{(\nu-\mu)/(\nu-\lambda)} \left(|u|_{1/\nu}\right)^{(\mu-\lambda)/(\nu-\lambda)}.$$

(viii) The case $\lambda < -1/n < 0 < \mu < \nu$: If we apply part (v) with $\nu = \mu$, we obtain that

$$|u|_{-n} \leq \gamma \left(|u|_{1/\lambda}\right)^{(n\mu+1)/n(\mu-\lambda)} \left(|u|_{1/\mu}\right)^{-(n\lambda+1)/n(\mu-\lambda)}.$$

But we have by part (vii) with $\lambda = -1/n$

$$|u|_{1/\mu} \leq \gamma (|u|_{-n})^{n(\nu-\mu)/(n\nu+1)} \left(|u|_{1/\nu}\right)^{(n\mu+1)/(n\nu+1)}.$$

Hence it follows that

$$|u|_{1/\mu} \leq \gamma \left(|u|_{1/\nu}\right)^{(n\mu+1)/(n\nu+1)} \Big(\left(|u|_{1/\lambda}\right)^{(n\mu+1)/n(\mu-\lambda)} \times \left(|u|_{1/\mu}\right)^{-(n\lambda+1)/n(\mu-\lambda)}\Big)^{-n(\nu-\mu)/(n\nu+1)},$$

so that

$$|u|_{1/\mu} \leq \gamma \left(|u|_{1/\lambda}\right)^{(\nu-\mu)/(\nu-\lambda)} \left(|u|_{1/\nu}\right)^{(\mu-\lambda)/(\nu-\lambda)}.$$

The proof of Proposition 2.7 is now complete. □

First we consider inequality (2.1) when $m = 1$ and $a = 1$:

Lemma 2.8. *If $n < p < \infty$, then there exists a constant $C = C(p, n) > 0$ such that*

$$|u|_{1-n/p,\infty} \leq C|u|_{1,p}, \quad u \in C_0^1(\mathbf{R}^n). \tag{2.15}$$

Proof. It suffices to show that

$$|u(x) - u(y)| \leq C|u|_{1,p}|x - y|^{1-n/p}, \quad x, y \in \mathbf{R}^n. \tag{2.15$'$}$$

We let

$$d = |x - y|,$$

and

$$G = \{z \in \mathbf{R}^n : |z - x| \leq d, |z - y| \leq d\}.$$

Then it follows that

$$\int_G |u(x) - u(y)|\, dz \leq \int_G |u(x) - u(z)|\, dz + \int_G |u(z) - u(y)|\, dz,$$

so that

$$cd^n|u(x)-u(y)| \le \int_G |u(x)-u(z)|\,dz + \int_G |u(z)-u(y)|\,dz \tag{2.16}$$

for some constant $c > 0$. Indeed, it suffices to note that the set G contains a ball $B_{d/2}$ of radius $d/2$ and that the volume of $B_{d/2}$ is equal to

$$\frac{\pi^{n/2}}{2^n\Gamma(n/2+1)}d^n.$$

We estimate each term on the right of inequality (2.16).

(a) We make the change of variables

$$z = x + \rho\sigma, \quad \rho > 0,\ \sigma \in S_1(0),$$

where $S_1(0)$ is the unit sphere about the origin. Then we have by Hölder's inequality

$$\begin{aligned}
&\int_G |u(x)-u(z)|dz \\
&\le \int_0^d \int_{S_1(0)} |u(x+\rho\sigma)-u(x)|\rho^{n-1}\,d\rho\,d\sigma \\
&= \int_0^d \int_{S_1(0)} \left|\int_0^\rho \frac{d}{dt}(u(x+t\sigma))\,dt\right| \rho^{n-1}\,d\rho\,d\sigma \\
&= \int_0^d \int_{S_1(0)} \left|\int_0^\rho \sum_{j=1}^n \sigma_j \frac{\partial u}{\partial x_j}(x+t\sigma)\,dt\right| \rho^{n-1}\,d\rho d\sigma \\
&\le \int_0^d \int_{S_1(0)} \int_0^\rho |\operatorname{grad} u(x+t\sigma)|\,dt\rho^{n-1}\,d\rho\,d\sigma \\
&= \int_0^d \left\{\int_{S_1(0)} \int_0^\rho |\operatorname{grad} u(x+t\sigma)|\, t^{(n-1)/p}\cdot t^{(1-n)/p}\,dt\,d\sigma\right\} \rho^{n-1}d\rho \\
&\le \int_0^d \left(\int_{S_1(0)} \int_0^\rho |\operatorname{grad} u(x+t\sigma)|^p\, t^{n-1}\,dt\,d\sigma\right)^{1/p} \\
&\qquad \times \left(\int_{S_1(0)} \int_0^\rho t^{\frac{1-n}{p-1}}\,dtd\sigma\right)^{(p-1)/p} \rho^{n-1}\,d\rho \\
&= \left(\frac{p-1}{p-n}\omega_n\right)^{(p-1)/p} \int_0^d \rho^{n-n/p}\left(\int_{|z-x|<\rho} |\operatorname{grad} u(z)|^p\,dz\right)^{1/p} d\rho \\
&\le \left(\frac{p-1}{p-n}\omega_n\right)^{(p-1)/p} \frac{d^{n+1-n/p}}{n+1-n/p}\left(\int_{\mathbf{R}^n} |\operatorname{grad} u(z)|^p\,dz\right)^{1/p},
\end{aligned}$$

where ω_n is the surface area of the unit sphere $S_1(0)$:

$$\omega_n = \frac{\pi^{n/2}}{2^n \Gamma(n/2+1)}.$$

This proves that

$$\int_G |u(x) - u(z)|dz \leq C'|u|_{1,p} d^{n-n/p+1}. \tag{2.17}$$

(b) Similarly, we have

$$\int_G |u(y) - u(z)|\, dz \leq C'|u|_{1,p} d^{n-n/p+1}. \tag{2.18}$$

Therefore, carrying inequalities (2.17) and (2.18) into inequality (2.16), we obtain that

$$c|u(x) - u(y)| \leq C'|u|_{1,p} d^{1-n/p},$$

so that

$$|u(x) - u(y)| \leq C|u|_{1,p}|x-y|^{1-n/p}, \quad x, y \in \mathbf{R}^n.$$

The proof of Lemma 2.8 is complete. □

Lemma 2.9. *Let $1 \leq p < n$. We have, for all functions $u \in C_0^1(\mathbf{R}^n)$,*

$$|u|_{0,np/(n-p)} \leq \frac{p(n-1)}{2(n-p)} \prod_{i=1}^{n} |D_i u|_{0,p}^{1/n}. \tag{2.19}$$

Proof. (i) The case $p = 1$:

$$|u|_{0,n/(n-1)} \leq \frac{1}{2} \prod_{i=1}^{n} |D_i u|_{0,1}^{1/n}. \tag{2.20}$$

Since we have for $1 \leq j \leq n$

$$\begin{aligned} u(x) &= \int_{-\infty}^{x_j} D_j u(x_1, \cdots, x_j, \cdots, x_n) dx_j \\ &= -\int_{x_j}^{\infty} D_j u(x_1, \cdots, x_j, \cdots, x_n) dx_j, \end{aligned}$$

it follows that

$$\begin{aligned} 2|u(x)| &\leq \int_{-\infty}^{x_j} |D_j u(x)| dx_j + \int_{x_j}^{\infty} |D_j u(x)| dx_j \\ &= \int_{-\infty}^{\infty} |D_j u(x)| dx_j. \end{aligned}$$

This gives that

$$\begin{aligned} (2|u(x)|)^{n/(n-1)} &= ((2|u(x)|)^n)^{1/(n-1)} \\ &\leq \prod_{j=1}^{n} \left(\int_{-\infty}^{\infty} |D_j u(x)| dx_j \right)^{1/(n-1)}. \end{aligned}$$

Here we need the following:

Lemma 2.10 (Gagliardo). *Let F_j be a function defined on $\mathbf{R}^{n-1}$ depending on the variables $x_1, \cdots, \hat{x}_j, \cdots, x_n$ $(1 \leq j \leq n)$ and belonging to the space $L^{n-1}(\mathbf{R}^{n-1})$. Then the function $F(x)$ on $\mathbf{R}^n$, defined by the formula*

$$F(x) = F_1(x_2, \cdots, x_n) \cdots F_j(x_1, \cdots, \hat{x}_j, \cdots, x_n) \cdots F_n(x_1, \cdots, x_{n-1}),$$
$$x = (x_1, x_2, \cdots, x_n) \in \mathbf{R}^n,$$

belongs to the space $L^1(\mathbf{R}^n)$ and satisfies the inequality

$$\int_{\mathbf{R}^n} |F(x)|\, dx \leq \prod_{j=1}^{n} \|F_j\|_{L^{n-1}(\mathbf{R}^{n-1})}.$$

Granting Lemma 2.10 for the moment, we shall prove inequality (2.20). Now, applying Lemma 2.10 with

$$F_j(x_1, \cdots, \hat{x}_j, \cdots, x_n) = \left(\int_{-\infty}^{\infty} |D_j u(x_1, \cdots, x_j, \cdots, x_n)|\, dx_j \right)^{1/(n-1)},$$

we obtain that

$$\int_{\mathbf{R}^n} (2|u(x)|)^{n/(n-1)} dx \leq \int_{\mathbf{R}^n} \left(\prod_{j=1}^{n} \left(\int_{-\infty}^{\infty} |D_j u(x)|\, dx_j \right)^{1/(n-1)} \right) dx$$
$$\leq \prod_{j=1}^{n} \left(\int_{\mathbf{R}^n} |D_j u(x)|\, dx \right)^{1/(n-1)}.$$

This proves inequality (2.20).

Proof of Lemma 2.10. We prove Lemma 2.10 by induction on the dimension n.

(a) The case $n = 2$: Assume that

$$F_1(x_2), F_2(x_1) \in L^1(\mathbf{R}).$$

Then we have by Fubini's theorem

$$\iint_{\mathbf{R}^2} |F(x_1, x_2)|\, dx_1 dx_2 = \iint_{\mathbf{R}^2} |F_1(x_2) F_2(x_1)|\, dx_1 dx_2$$
$$= \int_{\mathbf{R}} |F_1(x_2)| dx_2 \int_{\mathbf{R}} |F_2(x_1)| dx_1.$$

This proves the lemma for $n = 2$.

(b) Now we assume that the lemma has been established for $n - 1$.

By Hölder's inequality, it follows that

$$
\begin{aligned}
&\int_{\mathbf{R}} |F(x)| dx_1 \\
&= |F_1(\hat{x}_1)| \left(\int_{\mathbf{R}} |F_2(\hat{x}_2)||F_3(\hat{x}_3)| \cdots |F_n(\hat{x}_n)| dx_1 \right) \\
&\le |F_1(\hat{x}_1)| \left(\int_{\mathbf{R}} |F_2(\hat{x}_2)|^{n-1} dx_1 \right)^{1/(n-1)} \\
&\quad \times \left[\int_{\mathbf{R}} (|F_3(\hat{x}_3)| \cdots |F_n(\hat{x}_n)|)^{(n-1)/(n-2)} \, dx_1 \right]^{(n-2)/(n-1)} \\
&\le |F_1(\hat{x}_1)| \left(\int_{\mathbf{R}} |F_2(\hat{x}_2)|^{n-1} dx_1 \right)^{1/(n-1)} \\
&\quad \times \left[\left(\int_{\mathbf{R}} |F_3(\hat{x}_3)|^{(n-1)/(n-2) \cdot (n-2)} dx_1 \right)^{1/(n-2)} \right. \\
&\quad \times \left. \left(\int_{\mathbf{R}} (|F_4(\hat{x}_4)| \cdots |F_n(\hat{x}_n)|^{\frac{n-1}{n-2}\frac{n-2}{n-3}} \, dx_1 \right)^{(n-3)/(n-2)} \right]^{(n-2)/(n-1)} \\
&\le |F_1(\hat{x}_1)| \left(\int_{\mathbf{R}} |F_2(\hat{x}_2)|^{n-1} dx_1 \right)^{1/(n-1)} \left(\int_{\mathbf{R}} |F_3(\hat{x}_3)|^{n-1} dx_1 \right)^{1/(n-1)} \\
&\quad \times \left(\int_{\mathbf{R}} (|F_4(\hat{x}_4)| \cdots |F_n(\hat{x}_n)|)^{(n-1)/(n-3)} \, dx_1 \right)^{(n-3)/(n-1)} \\
&\vdots \\
&\le |F_1(\hat{x}_1)| \prod_{j=2}^{n} \left(\int_{\mathbf{R}} |F_j(\hat{x}_j)|^{n-1} dx_1 \right)^{1/(n-1)}
\end{aligned}
$$

Hence we have again by Hölder's inequality

(2.21)

$$
\begin{aligned}
&\int_{\mathbf{R}^n} |F(x)| dx_1 \cdot dx_2 \cdots dx_n \\
&\le \int_{\mathbf{R}^{n-1}} |F_1(\hat{x}_1)| \left(\prod_{j=2}^{n} \int_{\mathbf{R}} |F_j(\hat{x}_j)|^{n-1} dx_1 \right)^{1/(n-1)} dx_2 \cdots dx_n \\
&\le \left(\int_{\mathbf{R}^{n-1}} |F_1(\hat{x}_1)|^{n-1} \, dx_2 \cdots dx_n \right)^{1/(n-1)} \\
&\quad \times \left[\int_{\mathbf{R}^{n-1}} \prod_{j=2}^{n} \left(\int_{\mathbf{R}} |F_j(\hat{x}_j)|^{n-1} dx_1 \right)^{1/(n-2)} dx_2 \cdots dx_n \right]^{(n-2)/(n-1)}
\end{aligned}
$$

But we remark that the function

$$G_j(\hat{x}_j) = \left(\int_{\mathbf{R}} |F_j(\hat{x}_j)|^{n-1} dx_1 \right)^{1/(n-2)}$$

depends on the variables $x_2, \cdots, \hat{x}_j, \cdots, x_n$ $(2 \leq j \leq n)$ and belongs to $L^{n-2}(\mathbf{R}^{n-2})$. Hence, applying the induction hypothesis to the last part of inequality (2.21), we obtain that

$$\begin{aligned} &\int_{\mathbf{R}^{n-1}} \prod_{j=2}^{n} \left(\int_{\mathbf{R}} |F_j(\hat{x}_j)|^{n-1} dx_1 \right)^{1/(n-2)} dx_2 \cdots dx_n \\ &\leq \prod_{j=2}^{n} \left[\int_{\mathbf{R}^{n-1}} \left(\int_{\mathbf{R}} |F_j(\hat{x}_j)|^{n-1} dx_1 \right) dx_2 \cdots d\hat{x}_j \cdots dx_n \right]^{1/(n-2)} \\ &= \prod_{j=2}^{n} \left(\|F_j\|_{L^{n-1}(\mathbf{R}^{n-1})} \right)^{(n-1)/(n-2)}. \end{aligned} \tag{2.22}$$

Therefore, combining inequalities (2.21) and (2.22), we find that

$$\int_{\mathbf{R}^n} |F(x)|\, dx \leq \prod_{j=1}^{n} \|F_j\|_{L^{n-1}(\mathbf{R}^{n-1})}.$$

This completes the induction and hence the proof of Lemma 2.10. □

The proof of inequality (2.20) (inequality (2.19) with $p = 1$) is complete.

(ii) The case $1 < p < n$: If we let

$$v = |u|^{p(n-1)/(n-p)},$$

then we have

$$\begin{aligned} |u|_{0,np/(n-p)} &= \left(\int_{\mathbf{R}^n} |u(x)|^{np/(n-p)}\, dx \right)^{(n-p)/np} \\ &= \left(\int_{\mathbf{R}^n} |v(x)|^{n/(n-1)}\, dx \right)^{(n-p)/np} \\ &= \left(|v|_{0,n/(n-1)} \right)^{(n-p)/p(n-1)}, \end{aligned}$$

and also

$$v \in H^{1,1}(\mathbf{R}^n).$$

Indeed, it suffices to note that the function $|u|$ is absolutely continuous on $\mathbf{R}^n$ and so

$$D_j v = \frac{(n-1)p}{n-p} |u|^{n(p-1)/(n-p)} D_j(|u|) \quad \text{a.e. in } \mathbf{R}^n.$$

Further we remark that by the triangle inequality

$$|D_j|u|\,| \le |D_j u|,$$

so that

$$|D_j v| \le \frac{(n-1)p}{n-p}|u|^{n(p-1)/(n-p)}|D_j u| \quad \text{a.e. in } \mathbf{R}^n.$$

Since the space $C_0^1(\mathbf{R}^n)$ is dense in $H^{1,1}(\mathbf{R}^n)$, we can apply inequality (2.20) to the function v to obtain that

$$|v|_{0,n/(n-1)} \le \frac{1}{2}\prod_{j=1}^{n}\left(\int_{\mathbf{R}^n}|D_j v(x)|\,dx\right)^{1/n}. \tag{2.23}$$

But we have by Hölder's inequality

$$\begin{aligned}
&\int_{\mathbf{R}^n}|D_j v(x)|\,dx \\
&\le \frac{(n-1)p}{n-p}\int_{\mathbf{R}^n}|u(x)|^{n(p-1)/(n-p)}|D_j u|\,dx \\
&\le \frac{(n-1)p}{n-p}\left(\int_{\mathbf{R}^n}|u(x)|^{np/(n-p)}\,dx\right)^{(p-1)/p}\left(\int_{\mathbf{R}^n}|D_j u(x)|^p\,dx\right)^{1/p}.
\end{aligned} \tag{2.24}$$

Therefore, combining inequalities (2.23) and (2.24), we find that

$$\begin{aligned}
&\left(|u|_{0,np/(n-p)}\right)^{(n-1)p/(n-p)} \\
&= |v|_{0,np/(n-p)} \\
&\le \frac{1}{2}\frac{(n-1)p}{n-p}\left(\int_{\mathbf{R}^n}|u(x)|^{np/(n-p)}\,dx\right)^{(p-1)/p}\prod_{j=1}^{n}|D_j u|_{0,p}^{1/n} \\
&= \frac{1}{2}\frac{(n-1)p}{n-p}\prod_{j=1}^{n}|D_j u|_{0,p}^{1/n}\left(|u|_{0,np/(n-p)}\right)^{n(p-1)/(n-p)},
\end{aligned}$$

so that

$$|u|_{0,np/(n-p)} \le \frac{1}{2}\frac{p(n-1)}{n-p}\prod_{j=1}^{n}|D_j u|_{0,p}^{1/n}.$$

This completes the proof of inequality (2.19) and hence that of Lemma 2.9. □

Now we consider inequality (2.1) when $a = j/m$. It is convenient to begin with a straightforward one-dimensional interpolation inequality which typifies and provides a basis for the proof of the more general case (Lemma 2.14):

Lemma 2.11. *Let $1 \le p < \infty$, $1 \le q < \infty$, $2/r = 1/p + 1/q$ and let $I = [a, b]$ be a bounded, closed interval. Then we have, for all functions $u \in C^2(I)$,*

$$\left(\int_I |u'(x)|^r \, dx\right)^{1/r} \le |I|^{1+1/r-1/p} \left(\int_I |u''(x)|^p \, dx\right)^{1/p} + 8|I|^{-1-1/r+1/p} \left(\int_I |u(x)|^q \, dx\right)^{1/q}. \tag{2.25}$$

Here $|I| = b - a$, the length of the interval I.

Proof. We let

$$\alpha = \frac{b-a}{4},$$

and

$$\begin{cases} x_1 \in [a, a+\alpha], \\ x_2 \in [a+3\alpha, b]. \end{cases}$$

Then, by virtue of the mean value theorem, one can find a point $x_{12} \in (x_1, x_2)$ such that

$$u'(x_{12}) = \frac{u(x_1) - u(x_2)}{x_1 - x_2}.$$

Hence we have for any $x \in [a, b]$

$$\begin{aligned} u'(x) &= u'(x_{12}) + \int_{x_{12}}^{x} u''(y) \, dy \\ &= \frac{u(x_1) - u(x_2)}{x_1 - x_2} + \int_{x_{12}}^{x} u''(y) \, dy, \end{aligned}$$

so that

$$|u'(x)| \le \frac{|u(x_1)| + |u(x_2)|}{2\alpha} + \int_a^b |u''(y)| \, dy.$$

Integrating both sides from a to $a + \alpha$ with respect to x_1 and then from $a + 3\alpha$ to b with respect to x_2, we obtain that

$$\begin{aligned} \alpha^2 |u'(x)| &\le \frac{1}{2} \int_a^{a+\alpha} |u(x_1)| \, dx_1 + \frac{1}{2} \int_{a+3\alpha}^{b} |u(x_2)| \, dx_2 \\ &\quad + \alpha^2 \int_a^b |u''(y)| \, dy \\ &\le \frac{1}{2} \int_a^b |u(y)| \, dy + \alpha^2 \int_a^b |u''(y)| \, dy. \end{aligned}$$

Hence, by Hölder's inequality, it follows that

$$
\begin{aligned}
&\left(\int_a^b \left(\alpha^2 |u'(x)|\right)^r dx\right)^{1/r} \\
&\le \left(\int_a^b \left(\frac{1}{2}\int_a^b |u(y)|\,dy + \alpha^2 \int_a^b |u''(y)|\,dy\right)^r dx\right)^{1/r} \\
&= (b-a)^{1/r}\left(\frac{1}{2}\int_a^b |u(y)|\,dy + \alpha^2 \int_a^b |u''(y)|dy\right) \\
&\le \frac{(b-a)^{1/r}}{2}\left((b-a)^{1-1/q}\left(\int_a^b |u(y)|^q\,dy\right)^{1/q}\right) \\
&\quad + \alpha^2 (b-a)^{1/r}\left((b-a)^{1-1/p}\left(\int_a^b |u(y)|^p dy\right)^{1/p}\right) \\
&= \frac{1}{2}(b-a)^{1/r+1-1/q}\left(\int_a^b |u(y)|^q\,dy\right)^{1/q} \\
&\quad + \alpha^2 (b-a)^{1/r+1-1/p}\left(\int_a^b |u''(y)|^p\,dy\right)^{1/p}.
\end{aligned}
$$

Therefore, dividing both sides by $\alpha^2 = (b-a)^2/16$, we obtain that

$$
\begin{aligned}
\left(\int_a^b |u'(x)|^r\,dx\right)^{1/r} &\le 8(b-a)^{-1-1/r+1/p}\left(\int_a^b |u(y)|^q\,dy\right)^{1/q} \\
&\quad + (b-a)^{1+1/r-1/p}\left(\int_a^b |u''(y)|^p\,dy\right)^{1/p},
\end{aligned}
$$

since $1/q = 2/r - 1/p$.

The proof of Lemma 2.11 is complete. □

Lemma 2.12. *Let $1 \le p \le \infty$, $1 \le q \le \infty$ and $2/r = 1/p + 1/q$. Then we have, for all functions $u \in C^2_{(0)}([0,\infty))$,*

$$
\left(\int_0^\infty |u'(x)|^r dx\right)^{1/r} \le 4\sqrt{2}\left(\int_0^\infty |u''(x)|^p\,dx\right)^{1/2p}\left(\int_0^\infty |u(x)|^q\,dx\right)^{1/2q}.
$$

Here $C^2_{(0)}([0,\infty))$ is the subspace of $C^2([0,\infty))$ consisting of functions u with compact support in the interval $[0,\infty)$.

Proof. (i) The case $1 \le p, q < \infty$: For each function $u \in C^2_{(0)}([0,\infty))$, we can choose a positive integer $\ell = \ell(u)$ such that

$$
u(x) = 0 \quad \text{for all } x \ge \ell.
$$

If I is a bounded, closed interval in $[0,\infty)$, we introduce two functions $K_1(I)$ and $K_2(I)$ by the following formulas (cf. the right of inequality (2.25)):

$$K_1(I) = |I|^{1+1/r-1/p}\left(\int_I |u''(x)|^r\,dx\right)^{1/r};$$

$$K_2(I) = 8|I|^{-(1+1/r-1/p)}\left(\int_I |u(x)|^q\,dx\right)^{1/q}.$$

Let k be an arbitrary positive integer. Then we define a bounded, closed interval

$$I_1 = [0, a_1],$$

where the positive constant a_1 is chosen in such a way that

$$\begin{cases} a_1 = \frac{\ell}{k} & \text{if } K_1([0,\frac{\ell}{k}]) > K_2([0,\frac{\ell}{k}]), \\ a_1 > \frac{\ell}{k},\ K_1([0,a_1]) = K_2([0,a_1]) & \text{if } K_1([0,\frac{\ell}{k}]) \le K_2([0,\frac{\ell}{k}]). \end{cases}$$

Here we remark that $K_1([0,x])$ and $K_2([0,x])$ are continuous functions of x which enjoy the following properties:

$$K_1([0,x]) \uparrow \infty \quad \text{as } x \to \infty,$$

$$K_2([0,x]) \le \frac{8}{x}\left(\int_0^\infty |u(x)|^p\,dx\right)^{1/p} \longrightarrow 0 \quad \text{as } x \to \infty.$$

Similarly, we define a bounded, closed interval

$$I_2 = [a_1, a_2],$$

where the positive constant a_2 is chosen in such a way that

$$\begin{cases} a_2 = a_1 + \frac{\ell}{k} & \text{if } K_1([a_1,\frac{\ell}{k}]) > K_2([a_1,\frac{\ell}{k}]), \\ a_2 > a_1 + \frac{\ell}{k},\ K_1([a_1,a_2]) = K_2([a_1,a_2]) & \text{if } K_1([a_1,\frac{\ell}{k}]) \le K_2([a_1,\frac{\ell}{k}]). \end{cases}$$

Continuing this process, we have after j steps (at most k steps)

$$[0,\ell] \subset \bigcup_{i=1}^{j} I_i,$$

since the length of each interval $I_i = [a_{i-1}, a_i]$ is greater than ℓ/k.

But we have by Lemma 2.11

$$\left(\int_{I_i} |u'(x)|^r\,dx\right)^{1/r}$$
$$\le K_1([a_{i-1},a_i]) + K_2([a_{i-1},a_i])$$

$$\leq \begin{cases} 2K_1\left([a_{i-1}, a_{i-1}+\frac{\ell}{k}]\right) & \text{if} \quad a_i = a_{i-1}+\frac{\ell}{k}, \\ 2\sqrt{K_1\left([a_{i-1}, a_i]\right) K_2\left([a_{i-1}, a_i]\right)} & \text{if} \quad a_i > a_{i-1}+\frac{\ell}{k} \end{cases}$$

$$= \begin{cases} 2\left(\frac{\ell}{k}\right)^{1+1/r-1/p}\left(\int_{I_i} |u''(x)|^p\,dx\right)^{1/p} & \text{if} \quad a_i = a_{i-1}+\frac{\ell}{k}, \\ 4\sqrt{2}\left(\int_{I_i} |u''(x)|^p\,dx\right)^{1/2p}\left(\int_{I_i} |u''(x)|^q\,dx\right)^{1/2q} & \text{if} \quad a_i > a_{i-1}+\frac{\ell}{k}. \end{cases}$$

Since $r/2p+r/2q = 1$, it follows from an application of Hölder's inequality that

$$\begin{aligned} &\int_0^\infty |u'(x)|^r\,dx \\ &= \sum_{i=1}^{j} \int_{I_i} |u'(x)|^r\,dx \\ &\leq \sum_{i=1}^{j} 2^r \left(\frac{\ell}{k}\right)^{r+1-r/p} \left(\int_{I_i} |u''(x)|^p\,dx\right)^{r/p} \\ &\quad + (4\sqrt{2})^r \sum_{i=1}^{j} \left(\int_{I_i} |u''(x)|^p\,dx\right)^{r/2p} \left(\int_{I_i} |u(x)|^q\,dx\right)^{r/2q} \\ &\leq 2^r \left(\frac{\ell}{k}\right)^{r+1-r/p} k \left(\int_0^\infty |u''(x)|^p dx\right)^{r/p} \\ &\quad + (4\sqrt{2})^r \left(\sum_{i=1}^{j} \int_{I_i} |u''(x)|^p\,dx\right)^{r/2p} \left(\sum_{i=1}^{j} \int_{I_i} |u(x)|^q\,dx\right)^{r/2q} \\ &= 2^r\,\ell^{r+1-r/p}\,k^{r/p-r} \left(\int_0^\infty |u''(x)|^p\,dx\right)^{r/p} \\ &\quad + (4\sqrt{2})^r \left(\int_0^\infty |u''(x)|^p\,dx\right)^{r/2p} \left(\int_0^\infty |u(x)|^q\,dx\right)^{r/2q}. \end{aligned}$$

Therefore, letting $k \to \infty$, we obtain that

(2.26)

$$\int_0^\infty |u'(x)|^r\,dx \leq (4\sqrt{2})^r \left(\int_0^\infty |u''(x)|^p\,dx\right)^{r/2p} \left(\int_0^\infty |u(x)|^q\,dx\right)^{r/2q}.$$

(ii) The case $p = \infty$ or $q = \infty$ can be obtained by letting $p \to \infty$ or $q \to \infty$ in inequality (2.26), respectively.

The proof of Lemma 2.12 is complete. □

Corollary 2.13. *Let $1 \leq p \leq \infty$, $1 \leq q \leq \infty$ and $2/r = 1/p + 1/q$. Then we have, for all functions $u \in C_0^2(-\infty, \infty)$,*

$$\left(\int_{-\infty}^\infty |u'(x)|^r\,dx\right)^{1/r} \leq 4\sqrt{2} \left(\int_{-\infty}^\infty |u''(x)|^p\,dx\right)^{1/2p} \left(\int_{-\infty}^\infty |u(x)|^q\,dx\right)^{1/2q}.$$

Proof. We let

$$u_+ = u|_{[0,\infty)},$$
$$u_- = u|_{(-\infty,0]}.$$

Then, applying Lemma 2.12 to the functions u_+ and u_-, we obtain that

$$\begin{aligned}\int_{-\infty}^{\infty} |u'(x)|^r\,dx &= \int_{-\infty}^{0} |u'_-(x)|^r\,dx + \int_0^{\infty} |u'_+(x)|^r\,dx \\ &\le (4\sqrt{2})^r \left(\int_{-\infty}^{0} |u''_-(x)|^p\,dx\right)^{r/2p} \left(\int_{-\infty}^{0} |u_-(x)|^q\,dx\right)^{r/2q} \\ &\quad + (4\sqrt{2})^r \left(\int_0^{\infty} |u''_+(x)|^p\,dx\right)^{r/2p} \left(\int_0^{\infty} |u_+(x)|^q\,dx\right)^{r/2q} \\ &\le (4\sqrt{2})^r \left(\left(\int_{-\infty}^{0} |u''_-(x)|^p\,dx + \int_0^{\infty} |u''_+(x)|^p\,dx\right)^{r/2p}\right. \\ &\quad \times \left.\left(\int_{-\infty}^{0} |u_-(x)|^q\,dx + \int_0^{\infty} |u_+(x)|^q\,dx\right)^{r/2q}\right) \\ &= (4\sqrt{2})^r \left(\int_{-\infty}^{\infty} |u''(x)|^p\,dx\right)^{r/2p} \left(\int_{-\infty}^{\infty} |u(x)|^q\,dx\right)^{r/2q}.\end{aligned}$$

The proof of Corollary 2.13 is complete. □

Lemma 2.14. *Let $1 \le p \le \infty$, $1 \le q \le \infty$ and let j, m be integers such that $0 \le j < m$, and let*

$$\frac{1}{r} = \frac{j}{n} + \frac{j}{m}\left(\frac{1}{p} - \frac{m}{n}\right) + \left(1 - \frac{j}{m}\right)\frac{1}{q}.$$

If $m - j - n/p$ is not a non-negative integer, then we have, for all functions $u \in C_0^m(\mathbf{R}^n)$,

$$|u|_{j,r} \le C|u|_{m,p}^{j/m}|u|_{0,q}^{1-j/m}. \tag{2.27}$$

Here $C = C(m, n, p, q, j)$ is a positive constant.

Proof. Inequality (2.27) is obvious if $j = 0$.

(I) The case $j = 1$: By induction on m, we show that inequality (2.27) holds for all $m > 1$.

(I-1) The case $m = 2$: Applying Corollary 2.12 to the function $D_1 u(x_1, \cdot)$ ($u \in C_0^2(\mathbf{R}^n)$), we obtain that

$$\int_{-\infty}^{\infty} |D_1 u(x_1, x')|^r\,dx_1 \le (4\sqrt{2})^r \left(\int_{-\infty}^{\infty} |D_1^2 u(x_1, x')|^p\,dx_1\right)^{r/2p}$$

$$\times \left(\int_{-\infty}^{\infty} |u(x_1, x')|^q \, dx_1 \right)^{r/2q}$$

Hence, by Hölder's inequality, it follows that

$$\begin{aligned}
&\int_{\mathbf{R}^{n-1}} \int_{\mathbf{R}} |D_1 u(x_1, x')|^r \, dx_1 \, dx' \\
&\leq (4\sqrt{2})^r \int_{\mathbf{R}^{n-1}} \left(\int_{\mathbf{R}} |D_1^2 u(x_1, x')|^p \, dx_1 \right)^{r/2p} \\
&\qquad \times \left(\int_{\mathbf{R}} |u(x_1, x')|^q \, dx_1 \right)^{r/2q} dx' \\
&\leq (4\sqrt{2})^r \left(\int_{\mathbf{R}^n} |D_1^2 u(x)|^p \, dx \right)^{r/2p} \left(\int_{\mathbf{R}^n} |u(x)|^q \, dx \right)^{r/2q} \\
&\leq (4\sqrt{2})^r |u|_{2,p}^{r/2} |u|_{0,q}^{r/2},
\end{aligned}$$

so that

(2.28) $$\left(\int_{\mathbf{R}^n} |D_1 u(x)|^r \, dx \right)^{1/r} \leq 4\sqrt{2} |u|_{2,p}^{1/2} |u|_{0,q}^{1/2}.$$

Similarly, we have for $2 \leq j \leq n$

(2.28′) $$\left(\int_{\mathbf{R}^n} |D_j u(x)|^r \, dx \right)^{1/r} \leq 4\sqrt{2} |u|_{2,p}^{1/2} |u|_{0,q}^{1/2}.$$

Therefore, inequality (2.27) for $j = 1$ and $m = 2$ follows from inequalities (2.28) and (2.28′):

$(2.27)_{1,2}$ $$|u|_{1,r} \leq C |u|_{2,p}^{1/2} |u|_{0,q}^{1/2}, \quad u \in C_0^2(\mathbf{R}^n).$$

(I-2) Now we assume that inequality (2.27) has been established for $m-1$ (and $j = 1$):

$(2.27)_{1,m-1}$ $$|u|_{1,r'} \leq C |u|_{m-1,p}^{1/(m-1)} |u|_{0,q}^{1-1/(m-1)}, \quad u \in C_0^{m-1}(\mathbf{R}^n).$$

Then, applying inequality $(2.27)_{1,m-1}$ to the functions $D_j u$, $1 \leq j \leq n$ ($u \in C_0^m(\mathbf{R}^n)$), we obtain that

$$|D_j u|_{1,p_1} \leq C |D_j u|_{m-1,p}^{1/(m-1)} |D_j u|_{0,r}^{1-1/(m-1)},$$

so that

$$|u|_{2,p_1} \leq C |u|_{m,p}^{1/(m-1)} |u|_{1,r}^{1-1/(m-1)}.$$

But we have

$$|u|_{1,r} \leq C|u|_{2,p_1}^{1/2} |u|_{0,q}^{1/2}, \quad u \in C_0^2(\mathbf{R}^n).$$

Hence we have

$$\begin{aligned} |u|_{1,r} &\leq C \left(|u|_{m,p}^{1/(m-1)} |u|_{1,r}^{1-1/(m-1)} \right)^{1/2} |u|_{0,q}^{1/2} \\ &\leq C|u|_{m,p}^{1/2(m-1)} |u|_{1,r}^{(m-2)/2(m-1)} |u|_{0,q}^{1/2}. \end{aligned}$$

This proves inequality (2.27) for all $m > 1$ (and $j = 1$):

$$(2.27)_{1,m} \qquad |u|_{1,r} \leq C|u|_{m,p}^{1/m} |u|_{0,q}^{1-1/m}, \quad u \in C_0^m(\mathbf{R}^n).$$

(II) By induction on j, we show that inequality (2.27) holds for $0 < j < m$.

(II-1) Step (I) establishes the case $j = 1$.

(II-2) We assume that inequality (2.27) has been established for $j - 1$:

$$(2.27)_{j-1,\ell} \qquad |u|_{j-1,r'} \leq C|u|_{\ell,p}^{(j-1)/\ell} |u|_{0,q}^{1-(j-1)/\ell}, \quad \ell > j - 1.$$

Let m be an arbitrary integer such that

$$0 < j < m.$$

Since $m - 1 > j - 1$, one can apply inequality $(2.27)_{j-1,m-1}$ to the functions $D_j u$, $1 \leq j \leq n$ ($u \in C_0^m(\mathbf{R}^n)$), we obtain that

$$|D_j u|_{j-1,r} \leq C|D_j u|_{m-1,p}^{(j-1)/(m-1)} |D_j u|_{0,q_1}^{1-(j-1)/(m-1)},$$

so that

$$|u|_{j,r} \leq C|u|_{m,p}^{(j-1)/(m-1)} |u|_{1,q_1}^{1-(j-1)/(m-1)}.$$

But we have

$$|u|_{1,q_1} \leq C|u|_{m,p}^{1/m} |u|_{0,q}^{1-1/m}.$$

Hence it follows that

$$\begin{aligned} |u|_{j,r} &\leq C|u|_{m,p}^{(j-1)/(m-1)} \left(|u|_{m,p}^{1/m} |u|_{0,q}^{1-1/m} \right)^{1-(j-1)/(m-1)} \\ &\leq C|u|_{m,p}^{j/m} |u|_{0,q}^{1-j/m}. \end{aligned}$$

This completes the induction and hence the proof of Lemma 2.14. □

Proof of Theorem 2.1. (i) The case $a = j/m$:

$$|u|_{j,r} \le C|u|_{m,p}^{j/m}|u|_{0,q}^{1-j/m}.$$

This is nothing but inequality (2.27) in Lemma 2.14.

(ii) The case $a = 1$:

$$|D^j u|_r \le C|u|_{m,p}, \quad \frac{1}{r} = \frac{j}{n} + \frac{1}{p} - \frac{m}{n}. \tag{2.29}$$

Here we remark that

$$1 \le p < \infty,$$

since $m - j - n/p$ is not a non-negative integer.

(ii-1) The case $j = 0$:

$$|u|_r \le C|u|_{m,p}, \quad \frac{1}{r} = \frac{1}{p} - \frac{m}{n}. \tag{2.30}$$

(ii-1-a) The case $m = 1$:

$$|u|_r \le C|u|_{1,p}, \quad \frac{1}{r} = \frac{1}{p} - \frac{1}{n}. \tag{2.30$_1$}$$

Since $1 - n/p$ is not a non-negative integer, it follows that

$$p \ne n.$$

If $n < p < \infty$, then we have

$$r = \frac{np}{n-p} < 0.$$

Hence it follows from Lemma 2.8 that

$$|u|_r = |u|_{-n/r,\infty} = |u|_{1-n/p,\infty} \le C|u|_{1,p}.$$

If $1 \le p < n$, then we have

$$r > 0.$$

Hence it follows from Lemma 2.9 that

$$|u|_r = |u|_{0,np/(n-p)} \le \frac{p(n-1)}{2(n-p)}|u|_{1,p}.$$

Therefore, we have proved inequality $(2.30)_1$.

(ii-1-b) We assume that inequality (2.30) has been established for $m-1$:

$$(2.30)_{m-1} \qquad |u|_{r'} \leq C|u|_{m-1,p}, \quad \frac{1}{r'} = \frac{1}{p} - \frac{m-1}{n}.$$

We let

$$\frac{1}{r} = \frac{1}{p} - \frac{m}{n},$$
$$\frac{1}{r_1} = \frac{1}{r} + \frac{1}{n} = \frac{1}{p} - \frac{m-1}{n}.$$

Then, since $m-1-n/p$ is not a non-negative integer, we can apply inequality $(2.30)_{m-1}$ with $r' = r_1$ to the functions $D_i u$, $1 \leq i \leq n$, to obtain that

$$(2.31) \qquad |D_i u|_{r_1} \leq C|D_i u|_{m-1,p} \leq C|u|_{m,p}.$$

If $r < 0$, then we have

$$1 \leq p < r_1 < r.$$

Hence, by inequality $(2.30)_1$ with $p = r_1$, it follows that

$$(2.32) \qquad |u|_r \leq C|u|_{1,r_1}.$$

Thus, combining inequalities (2.32) and (2.31), we obtain that

$$|u|_r \leq C|u|_{1,r_1} \leq C|u|_{m,p}.$$

If $r < 0$, then we have

$$r_1 > 0 \implies r_1 > n,$$
$$r_1 < 0 \implies -\frac{n}{r_1} = \left[-\frac{n}{r}\right] - 1 + h, \quad h = -\frac{n}{r} - \left[-\frac{n}{r}\right].$$

Hence, if $r_1 > 0$, we have by Lemma 2.8

$$|u|_r = |u|_{-n/r,\infty} = |u|_{1-n/r_1,\infty} \leq C|u|_{1,r_1}.$$

Combining this inequality with inequality (2.31), we obtain that

$$|u|_r \leq C|u|_{m,p}.$$

If $r_1 < 0$, then we have

$$|u|_r = |u|_{-n/r,\infty}$$
$$= \sup_{x \neq y} \frac{|D^\ell u(x) - D^\ell u(y)|}{|x-y|^h}$$

$$= \sup_{\substack{x \neq y \\ 1 \leq i \leq n}} \frac{|D^{\ell-1} D_i u(x) - D^{\ell-1} D_i u(y)|}{|x-y|^h}$$
$$\leq \max_{1 \leq i \leq n} |D_i u|_{r_1}.$$

Therefore, combining this inequality with inequality (2.31), we find that

$$|u|_r \leq C|u|_{m,p}.$$

(ii-2) We assume that inequality (2.29) has been established for $j-1$:

$$|D^{j-1}u|_{r'} \leq C|u|_{m,p}, \quad \frac{1}{r'} = \frac{j-1}{n} + \frac{1}{p} - \frac{m}{n}.$$

We let

$$\frac{1}{r} = \frac{j}{n} + \frac{1}{p} - \frac{m}{n}.$$

Then we remark that

$$\frac{1}{r} = \frac{j-1}{n} + \frac{1}{p} - \frac{m-1}{n},$$

and that

$$(m-1) - (j-1) - n/p \text{ is not a non-negative integer.}$$

Hence we have by the induction hypothesis

$$|D^{j-1}v|_r \leq C|v|_{m-1,p}, \quad v \in C_0^{m-1}(\mathbf{R}^n).$$

Therefore, applying this inequality to the functions $D_i u$, $1 \leq i \leq n$, we obtain that

$$|D^j u|_r = |D^{j-1}(Du)|_r \leq C|Du|_{m-1,p} \leq C|u|_{m,p}.$$

This completes the proof of inequality (2.29) (inequality (2.1) with $a = 1$).

(iii) The case $j/m < a < 1$: We let

$$\frac{1}{r} = \frac{j}{n} + a\left(\frac{1}{p} - \frac{m}{n}\right) + (1-a)\frac{1}{q},$$
$$\frac{1}{r_1} = \frac{j}{m}\frac{1}{p} + \left(1 - \frac{j}{m}\right)\frac{1}{q},$$
$$\frac{1}{r_2} = \frac{j}{n} + \frac{1}{p} - \frac{m}{n},$$

and

$$\theta = \frac{1-a}{1-j/m}.$$

Then we have by parts (i) and (ii)

$$|D^j u|_{r_1} \le C|u|_{m,p}^{j/m}|u|_q^{1-j/m},$$
$$|D^j u|_{r_2} \le C|u|_{m,p}.$$

But we remark that

$$\frac{1}{r} = \frac{\theta}{r_1} + \frac{1-\theta}{r_2}, \quad 0 < \theta < 1.$$

Therefore, applying Proposition 2.7 to our situation, we find that

$$\begin{aligned}|D^j u|_{r_1} &\le C|D^j u|_{r_1}^{\theta}|D^j u|_{r_2}^{1-\theta} \\ &\le C\left(|u|_{m,p}^{j/m}|u|_q^{1-j/m}\right)^{(1-a)/(1-j/m)}\left(|u|_{m,p}\right)^{(a-j/m)/(1-j/m)} \\ &= C|u|_{m,p}^{a}|u|_q^{1-a}.\end{aligned}$$

This completes the proof of Theorem 2.1. □

2.3 Imbeddings of the Spaces $H^{m,p}(\mathbf{R}^n)$

It is the imbedding characteristics of Sobolev spaces that render these spaces so useful in the study of partial differential equations. The most important of the imbedding properties of the spaces $H^{m,p}(\mathbf{R}^n)$ are usually lumped together in a single theorem referred to as the *Sobolev imbedding theorem.*

Theorem 2.15. *Let $1 \le p \le \infty$, $1 \le q \le \infty$ and let j, m be integers such that $0 \le j < m$. If $m - j - n/p$ is not a non-negative integer and if*

$$\frac{1}{r} = \frac{j}{n} + a\left(\frac{1}{p} - \frac{m}{n}\right) + (1-a)\frac{1}{q}, \quad \frac{j}{m} \le a \le 1,$$

then we have, for all functions $u \in H^{m,p}(\mathbf{R}^n) \cap L^q(\mathbf{R}^n)$,

$$|D^j u|_r \le C|u|_{m,p}^{a}|u|_{0,q}^{1-a}. \tag{2.33}$$

Here $C = C(m,n,p,q,j,a)$ is a positive constant.

Proof. (1) First we show the following:

Lemma 2.16. *If $u \in H^{m,p}(\mathbf{R}^n) \cap L^q(\mathbf{R}^n)$, then one can find a sequence $\{u_\nu\} \subset C_0^m(\mathbf{R}^n)$ such that*

$$u_\nu \longrightarrow u \quad \text{in } H^{m,p}(\mathbf{R}^n), \tag{2.34}$$
$$u_\nu \longrightarrow u \quad \text{in } L^q(\mathbf{R}^n). \tag{2.35}$$

Proof. Take a function $\rho \in C_0^\infty(\mathbf{R}^n)$ such that

$$\begin{cases} \operatorname{supp} \rho \subset \{x \in \mathbf{R}^n : |x| \leq 1\}, \\ \int_{\mathbf{R}^n} \rho(x)\, dx = 1, \end{cases}$$

and define

$$\rho_\varepsilon(x) = \frac{1}{\varepsilon^n} \rho\left(\frac{x}{\varepsilon}\right), \quad \varepsilon > 0.$$

Then we have

$$u_\varepsilon(x) = u * \rho_\varepsilon(x) = \int_{\mathbf{R}^n} \rho_\varepsilon(x-y) u(y)\, dy \in C^\infty(\mathbf{R}^n),$$

and

$$\begin{aligned} u_\varepsilon &\longrightarrow u \quad \text{in } H^{m,p}(\mathbf{R}^n) \text{ as } \varepsilon \downarrow 0, \\ u_\varepsilon &\longrightarrow u \quad \text{in } L^q(\mathbf{R}^n) \text{ as } \varepsilon \downarrow 0. \end{aligned}$$

Furthermore, take a function $\varphi \in C_0^m(\mathbf{R}^n)$ such that

$$\varphi(x) = \begin{cases} 1 & \text{if } |x| \leq 1, \\ 0 & \text{if } |x| \geq 2, \end{cases}$$

and define

$$\varphi_R(x) = \varphi\left(\frac{x}{R}\right), \quad R > 0.$$

Then it follows that

$$\varphi_R\, u_\varepsilon \longrightarrow u \quad \text{in } H^{m,p}(\mathbf{R}^n) \text{ as } R \to \infty \text{ and } \varepsilon \downarrow 0, \tag{2.34'}$$

$$\varphi_R\, u_\varepsilon \longrightarrow u \quad \text{in } L^q(\mathbf{R}^n) \text{ as } R \to \infty \text{ and } \varepsilon \downarrow 0. \tag{2.35'}$$

This proves the claim, since $\{\varphi_R\, u_\varepsilon\} \subset C_0^m(\mathbf{R}^n)$.

Indeed, we have for $|\alpha| \leq m$

(2.36)

$$\int_{\mathbf{R}^n} |D^\alpha(\varphi_R\, u_\varepsilon) - D^\alpha u|^p\, dx \leq 2^{p-1}\biggl(\int_{\mathbf{R}^n} |D^\alpha(\varphi_R\, u_\varepsilon) - D^\alpha(\varphi_R\, u)|^p\, dx + \int_{\mathbf{R}^n} |D^\alpha(\varphi_R\, u) - D^\alpha u|^p\, dx\biggr).$$

But, since we have

$$\varphi_R(x) = 1, \quad |x| \leq R,$$

$$D^\gamma(\varphi_R(x)) = D^\alpha \varphi\left(\frac{x}{R}\right) \cdot \left(\frac{1}{R}\right)^{|\gamma|},$$

it follows that, as $R \to \infty$,

(2.37)
$$\begin{aligned}\int_{\mathbf{R}^n} |D^\alpha(\varphi_R u) - D^\alpha u|^p \, dx &= \int_{|x|>R} |D^\alpha(\varphi_R u) - D^\alpha u|^p \, dx \\ &= \int_{|x|>R} \left| \sum_{\beta \le \alpha} \binom{\alpha}{\beta} D^{\alpha-\beta}\varphi_R \cdot D^\beta u - D^\alpha u \right| dx \\ &\le C \sum_{|\gamma| \le m} \int_{|x|>R} |D^\gamma u|^p \, dx \longrightarrow 0.\end{aligned}$$

Further, we have, as $\varepsilon \downarrow 0$,

(2.38)
$$\begin{aligned}&\int_{\mathbf{R}^n} |D^\alpha(\varphi_R u_\varepsilon) - D^\alpha(\varphi_R u)|^p \, dx \\ &= \int_{\mathbf{R}^n} |D^\alpha(\varphi_R(u_\varepsilon - u))|^p \, dx \\ &= \int_{\mathbf{R}^n} \left| \sum_{\beta \le \alpha} \binom{\alpha}{\beta} D^{\alpha-\beta}\varphi_R \cdot D^\beta(u_\varepsilon - u) \right|^p dx \\ &\le C \sum_{|\beta| \le m} \int_{\mathbf{R}^n} \left| D^\beta(u_\varepsilon - u) \right|^p dx \longrightarrow 0.\end{aligned}$$

Therefore, assertion (2.34′) follows from inequalities (2.36), (2.37) and (2.38).

Assertion (2.35′) is proved similarly.

The proof of Lemma 2.16 is complete. □

(2) By Lemma 2.16, one can find a sequence $\{u_\nu\} \subset C_0^m(\mathbf{R}^n)$ such that

$$u_\nu \longrightarrow u \quad \text{in } H^{m,p}(\mathbf{R}^n),$$
$$u_\nu \longrightarrow u \quad \text{in } L^q(\mathbf{R}^n).$$

Hence, applying inequality (2.1) to the functions u_ν and $u_\nu - u_\mu$, we obtain that

$$|D^j u_\nu|_r \le C |u_\nu|_{m,p}^a |u_\nu|_{0,q}^{1-a}, \tag{2.39}$$
$$|D^j u_\nu - D^j u_\mu|_r \le C |u_\nu - u_\mu|_{m,p}^a |u_\nu - u_\mu|_{0,q}^{1-a}. \tag{2.40}$$

(2-a) If $r \ge 1$, then it follows that

$$D^j u_\nu \longrightarrow D^j u \quad \text{in } L^r(\mathbf{R}^n).$$

Thus, passing to the limit in inequality (2.39), we find that

$$|D^j u|_r \le C |u|_{m,p}^a |u|_{0,q}^{1-a}.$$

(2-b) If $r < 0$, then we let

$$\ell = \left[-\frac{n}{r}\right],$$

so that

$$j + \ell \leq j - \frac{n}{r} = am - a\frac{n}{p} - (1-a)\frac{n}{q} < m.$$

One may assume that for $|\alpha| \leq m$

$$D^\alpha u_\nu(x) \longrightarrow D^\alpha u(x) \quad \text{a.e. } x \in \mathbf{R}^n. \tag{2.41}$$

(2-b-i) The case $-n/r \in \mathbf{N}$: Then we have by inequality (2.39)

$$|D^{j+\ell} u_\nu|_{0,\infty} = |D^j u_\nu|_{\ell,\infty} = |D^j u_\nu|_r \leq C |u_\nu|_{m,p}^a |u_\nu|_{0,q}^{1-a}. \tag{2.42}$$

This implies that the sequence $\{D^{j+\ell}u_\nu\}$ is uniformly bounded on $\mathbf{R}^n$. Thus, by the mean value theorem, it follows that the sequence $\{D^\alpha u_\nu\}$ is equicontinuous on every bounded subset of $\mathbf{R}^n$, for $|\alpha| = j + \ell - 1$.

Here we need the following elementary lemma:

Lemma 2.17. *Let $\{u_\nu\}_{\nu=1}^\infty$ be a sequence in the space $C^m(\mathbf{R}^n)$. Assume that:*

(i) There is a point $x_0 \in \mathbf{R}^n$ such that the set

$$\{D^\alpha u_\nu(x_0) : \nu = 1, 2, \cdots, \ |\alpha| \leq m\}$$

is uniformly bounded, that is, for some constant $M > 0$

$$\sup_{|\alpha| \leq m} \sup_\nu |D^\alpha u_\nu(x_0)| \leq M.$$

(ii) The set $\{D^\alpha u_\nu : \nu = 1, 2, \cdots, |\alpha| = m\}$ of functions is equicontinuous on every bounded subset of $\mathbf{R}^n$.

Then it follows that the set $\{D^\alpha u_\nu\}$ is uniformly bounded and equicontinuous on every bounded subset of $\mathbf{R}^n$, for $|\alpha| \leq m$.

Proof. First, conditions (i) and (ii) imply that

(a) The set $\{D^\alpha u_\nu\}$ is uniformly bounded on every bounded subset of $\mathbf{R}^n$, for $|\alpha| = m$.

Hence, by the mean value theorem, it follows that

(b) The set $\{D^\beta u_\nu\}$ is equicontinuous on every bounded subset of $\mathbf{R}^n$, for $|\beta| = m - 1$.

Therefore, this together with condition (i) implies that

(c) The set $\{D^\beta u_\nu\}$ is uniformly bounded on every bounded subset of $\mathbf{R}^n$, for $|\beta| = m - 1$.

Lemma 2.17 follows by repeating this process. □

Now we find from Lemma 2.17 that the sequence $\{D^\alpha u_\nu\}$ is uniformly bounded and equicontinuous on every bounded subset of $\mathbf{R}^n$, for $|\alpha| \leq j + \ell - 1$. By the Ascoli-Arzelà theorem, it follows that

$$u \in C^{j+\ell-1}(\mathbf{R}^n),$$
$$u_\nu \longrightarrow u \quad \text{in } C^{j+\ell-1}(\mathbf{R}^n).$$

But we have by inequality (2.40)

$$\begin{aligned} |D^{j+\ell}u_\nu - D^{j+\ell}u_\mu|_{0,\infty} &= |D^j u_\nu - D^j u_\mu|_{\ell,\infty} \\ &= |D^j u_\nu - D^j u_\mu|_r \\ &\leq C|u_\nu - u_\mu|_{m,p}^a |u_\nu - u_\mu|_{0,q}^{1-a}. \end{aligned}$$

Therefore, we obtain that

$$u \in C^{j+\ell}(\mathbf{R}^n),$$
$$|D^{j+\ell}u_\nu - D^{j+\ell}u|_{0,\infty} \longrightarrow 0 \quad \text{as } \nu \to \infty.$$

Further, by passing to the limit in inequality (2.42), we have

$$|D^j u|_r = |D^{j+\ell}u|_{0,\infty} \leq C|u|_{m,p}^a |u|_{0,q}^{1-a}.$$

(2-b-ii) The case $-n/r \notin \mathbf{N}$: We let

$$-\frac{n}{r} = \ell + h, \quad 0 < h < 1.$$

Then we have by inequality (2.39)

$$|D^{j+\ell}u_\nu(x) - D^{j+\ell}u_\nu(y)| \leq C|u_\nu|_{m,p}^a |u_\nu|_{0,q}^{1-a} |x-y|^h. \tag{2.43}$$

In view of (2.41), it follows from an application of Lemma 2.17 that the sequence $\{D^\alpha u_\nu\}$ is uniformly bounded and equicontinuous on every bounded subset of $\mathbf{R}^n$, for $|\alpha| \leq j + \ell$. Hence we find from the Ascoli-Arzelà theorem that

$$u \in C^{j+\ell}(\mathbf{R}^n),$$
$$u_\nu \longrightarrow u \quad \text{in } C^{j+\ell}(\mathbf{R}^n).$$

By passing to the limit in inequality (2.43), we obtain that

$$|D^{j+\ell}u(x) - D^{j+\ell}u(y)| \leq C|u|_{m,p}^a |u|_{0,q}^{1-a} |x-y|^h,$$

so that

$$|D^j u|_r = |D^j u|_{-n/r,\infty} = |D^{j+\ell}u|_{h,\infty} \leq C|u|_{m,p}^a |u|_{0,q}^{1-a}.$$

The proof of Theorem 2.15 is now complete. □

The next imbedding theorem asserts the existence of imbeddings of the spaces $H^{m,p}(\mathbf{R}^n)$ into the spaces $B^a(\mathbf{R}^n)$:

Theorem 2.18. *Let $1 \le p < \infty$ and let m be a positive integer such that $m > n/p$. If n/p is not an integer, then we have*

$$H^{m,p}(\mathbf{R}^n) \subset B^{m-n/p}(\mathbf{R}^n),$$

with continuous injection. Moreover, we have

$$|u|_{m-n/p,\infty} \le \gamma |u|_{m,p}, \tag{2.44}$$

$$\|u\|_{m-[n/p]-1,\infty} \le C\|u\|_{m,p}. \tag{2.45}$$

Remark 2.19. Since elements of $H^{m,p}(\mathbf{R}^n)$ are not functions defined everywhere on $\mathbf{R}^n$ but equivalence classes of such functions defined and equal up to sets of measure zero, we must clarify what is meant by the imbedding $H^{m,p}(\mathbf{R}^n)$ into $B^{m-n/p}(\mathbf{R}^n)$. What is intended is that each $u \in H^{m,p}(\mathbf{R}^n)$ can be redefined on a set of measure zero in $\mathbf{R}^n$ in such a way that the modified function $\tilde{u}$ (which equals u in $H^{m,p}(\mathbf{R}^n)$) belongs to $B^{m-n/p}(\mathbf{R}^n)$.

The proof of Theorem 2.18 is based on the following lemma:

Lemma 2.20. *Let $1 \le p < \infty$ and let j, m be integers such that $0 \le j \le m - n/p$. Then we have, for all functions $u \in C_0^m(\mathbf{R}^n)$,*

$$|u|_{j,\infty} \le C|u|_{m,p}^{(n+jp)/mp}|u|_{0,p}^{1-(n+jp)/mp}. \tag{2.46}$$

Here $C = C(m,p,n)$ is a positive constant.

Proof. (1) We let

$$\ell = m - j,$$

and so

$$\frac{n}{p} < \ell.$$

Then, to prove inequality (2.46), it suffices to show the following:

$$|u|_{0,\infty} \le C|u|_{\ell,p}^{n/\ell p}|u|_{0,p}^{1-n/\ell p}, \quad u \in C_0^\ell(\mathbf{R}^n). \tag{2.47}$$

Indeed, applying inequality (2.47) to the functions

$$D^\alpha u, \quad |\alpha| = j, \ u \in C_0^m(\mathbf{R}^n),$$

we obtain that

$$|D^\alpha u|_{0,\infty} \le C|D^\alpha u|_{\ell,p}^{n/\ell p}|D^\alpha u|_{0,p}^{1-n/\ell p},$$

so that

$$|u|_{j,\infty} \le C|u|_{m,p}^{\ell p}|u|_{j,p}^{1-n/\ell p}. \tag{2.48}$$

But we have by Lemma 2.14

$$|u|_{j,p} \leq C|u|_{m,p}^{j/m}|u|_{0,p}^{1-j/m}.$$

Therefore, combining this inequality with inequality (2.48), we find that

$$\begin{aligned}|u|_{j,\infty} &\leq C|u|_{m,p}^{n/\ell p}\left(|u|_{m,p}^{j/m}|u|_{0,p}^{1-j/p}\right)^{1-n\ell/p} \\ &\leq C|u|_{m,p}^{(n+jp)/mp}|u|_{0,p}^{1-(n+jp)/mp}.\end{aligned}$$

(2) *Proof of inequality (2.47).* It suffices to show that

There exists a constant $C = C(\ell, p, n) > 0$ such that

$$|u(x)| \leq C|u|_{\ell,p}^{n/\ell p}|u|_{0,p}^{1-n/\ell p}, \quad x \in \mathbf{R}^n. \tag{2.47$'$}$$

Without loss of generality, we may assume that

$$x = 0.$$

If $u \in C_0^\ell(\mathbf{R}^n)$, we let

$$f(t) = u(t\sigma), \quad t > 0,\ \sigma \in \Sigma,$$

where Σ is the unit sphere. Then, applying Taylor's formula to the function f, we obtain that

$$\begin{aligned}f(t_1) = f(t_2) + f'(t_2)(t_1 - t_2) + \cdots + \frac{f^{(\ell-1)}(t_2)}{(\ell-1)!}(t_1 - t_2)^{\ell-1} \\ + \int_{t_1}^{t_2} \frac{(t_1 - \tau)^{\ell-1}}{(\ell-1)!} f^{(\ell)}(\tau)\, d\tau.\end{aligned}$$

If we take

$$t_1 = 0,\ t_2 = t,$$

it follows that

$$u(0) = \sum_{k=0}^{\ell-1} \frac{(-t)^k}{k!}\left(\frac{d}{dt}\right)^k (u(t\sigma)) - \int_0^t \frac{(-\tau)^{\ell-1}}{(\ell-1)!}\left(\frac{d}{d\tau}\right)^\ell (u(\tau\sigma))\, d\tau.$$

Hence, integrating both sides over the ball of radius h, we obtain that

$$\begin{aligned}&|u(0)|\left(\frac{h^n}{n}\omega_n\right)^{1/p} \\ &= \left(\int_0^h \int_\Sigma |u(0)|^p t^{n-1}\, dt\, d\sigma\right)^{1/p} \\ &\leq \sum_{k=0}^{\ell-1} \frac{1}{k!}\left(\int_0^h \int_\Sigma t^{kp}\left|\left(\frac{d}{dt}\right)^k (u(t\sigma))\right|^p t^{n-1}\, dt\, d\sigma\right)^{1/p} \\ &\quad + \frac{1}{(\ell-1)!}\left(\int_0^h \int_\Sigma \left|\int_0^t \tau^{\ell-1}\left(\frac{d}{d\tau}\right)^\ell (u(\tau\sigma))\, d\tau\right|^p t^{n-1}\, dt\, d\sigma\right)^{1/p}.\end{aligned} \tag{2.49}$$

Here

$$\omega_n = \frac{n\pi^{n/2}}{\Gamma(n/2+1)}, \qquad \text{the surface area of the unit ball } \Sigma.$$

We estimate each term on the right of inequality (2.49).

(a)
$$\sum_{k=0}^{\ell-1} \frac{1}{k!} \left(\int_0^h \int_\Sigma t^{kp} \left| \left(\frac{d}{dt}\right)^k (u(t\sigma)) \right|^p t^{n-1}\, dt\, d\sigma \right)^{1/p}$$
$$\leq C \sum_{k=0}^{\ell-1} h^k \left(\int_0^h \int_\Sigma \left| \left(\frac{d}{dt}\right)^k (u(t\sigma)) \right|^p t^{n-1}\, dt\, d\sigma \right)^{1/p}$$
$$\leq C \sum_{k=0}^{\ell-1} h^k |u|_{k,p}.$$

(b) If $p > 1$, we have by Hölder's inequality

$$\left| \int_0^t \tau^{\ell-1} \left(\frac{d}{d\tau}\right)^\ell (u(\tau\sigma))\, d\tau \right|^p$$
$$= \left| \int_0^t \tau^{\ell-1-(n-1)/p} \tau^{(n-1)/p} \left(\frac{d}{d\tau}\right)^\ell (u(\tau\sigma))\, d\tau \right|^p$$
$$\leq \left(\int_0^t \tau^{(\ell-1-(n-1)/p)q}\, d\tau \right)^{p/q} \int_0^t \tau^{n-1} \left| \left(\frac{d}{d\tau}\right)^\ell (u(\tau\sigma)) \right|^p d\tau$$
$$= \left(\frac{p-1}{\ell p - n} \right)^{p-1} t^{\ell p - n} \int_0^t \tau^{n-1} \left| \left(\frac{d}{d\tau}\right)^\ell (u(\tau\sigma)) \right|^p d\tau.$$

If $p = 1$, we have

$$\left| \int_0^t \tau^{\ell-1} \left(\frac{d}{d\tau}\right)^\ell (u(\tau\sigma))\, d\tau \right| \leq t^{\ell-n} \int_0^t \tau^{n-1} \left| \left(\frac{d}{d\tau}\right)^\ell (u(\tau\sigma)) \right| d\tau,$$

so that the number

$$\left(\frac{p-1}{\ell p - n} \right)^{p-1}$$

should be replaced by 1 if $p = 1$.

Hence it follows that

$$\left(\int_0^h \int_\Sigma \left| \int_0^t \tau^{\ell-1} \left(\frac{d}{d\tau}\right)^\ell (u(\tau\sigma))\, d\tau \right|^p t^{n-1} dt d\sigma \right)^{1/p}$$

$$\leq \left[\int_0^h \int_\Sigma \left(\frac{p-1}{\ell p - n}\right)^{p-1} t^{\ell p - n} \times \left(\int_0^t \tau^{n-1} \left|\left(\frac{d}{d\tau}\right)^\ell (u(\tau\sigma))\right|^p d\tau\right) t^{n-1} dt d\sigma\right]^{1/p}$$

$$\leq C\left[\int_0^h \int_\Sigma t^{\ell p-1} \left(\int_0^t \tau^{n-1} \left|\left(\frac{d}{d\tau}\right)^\ell (u(\tau\sigma))\right|^p d\tau\right) dt d\sigma\right]^{1/p}.$$

But we have by integration by parts

$$\int_0^h t^{\ell p-1} \left(\int_0^t \tau^{n-1} \left|\left(\frac{d}{d\tau}\right)^\ell (u(\tau\sigma))\right|^p d\tau\right) dt$$

$$= \left[\frac{t^{\ell p}}{\ell p} \int_0^t \tau^{n-1} \left|\left(\frac{d}{d\tau}\right)^\ell (u(\tau\sigma))\right|^p d\tau\right]_{t=0}^{t=h} - \int_0^t \frac{t^{\ell p}}{\ell p} t^{n-1} \left|\left(\frac{d}{d\tau}\right)^\ell (u(t\sigma))\right|^p dt$$

$$\leq \frac{h^{\ell p}}{\ell p} \int_0^h \tau^{n-1} \left|\left(\frac{d}{d\tau}\right)^\ell (u(\tau\sigma))\right|^p d\tau.$$

Thus it follows that

$$\left(\int_0^h \int_\Sigma \left|\int_0^t \tau^{\ell-1} \left(\frac{d}{d\tau}\right)^\ell (u(\tau\sigma))\, d\tau\right|^p t^{n-1}\, dt\, d\sigma\right)^{1/p}$$

$$\leq C h^\ell \left(\int_\Sigma \int_0^h \tau^{n-1} \left|\left(\frac{d}{d\tau}\right)^\ell (u(\tau\sigma))\right|^p d\tau\, d\sigma\right)^{1/p}$$

$$\leq C h^\ell |u|_{\ell,p}.$$

Summing up, we have proved that

$$|u(0)| \left(\frac{h^n \omega_n}{n}\right)^{1/p} \leq C \sum_{k=0}^{\ell} h^k |u|_{k,p},$$

that is,

$$|u(0)| h^{n/p} \leq C \sum_{k=0}^{\ell} h^k |u|_{k,p}.$$

Therefore, combining this inequality with inequality (2.27) with $j = k$ and $m = \ell$, we find that

$$|u(0)| h^{n/p} \leq C \sum_{k=0}^{\ell} h^k |u|_{\ell,p}^{k/\ell} |u|_{0,p}^{1-k/\ell}$$

$$
\begin{aligned}
&= C \sum_{k=0}^{\ell} \left(h^\ell |u|_{\ell,p}\right)^{k/\ell} |u|_{0,p}^{1-k/\ell} \\
&\leq C \sum_{k=0}^{\ell} \left(\frac{k}{\ell} h^\ell |u|_{\ell,p} + \left(1 - \frac{k}{\ell}\right) |u|_{0,p}\right) \\
&\leq C \left(h^\ell |u|_{\ell,p} + |u|_{0,p}\right),
\end{aligned}
$$

so that

$$
|u(0)| \leq C \left(h^{\ell - n/p} |u|_{\ell,p} + h^{-n/p} |u|_{0,p}\right), \quad \text{for all } h > 0. \tag{2.50}
$$

Now we choose a number h_0 so that the function

$$
h^{\ell - n/p} |u|_{\ell,p} + h^{-n/p} |u|_{0,p}
$$

attains its minimum at $h = h_0$, that is,

$$
h_0 = \left(\frac{n|u|_{0,p}}{p|u|_{\ell,p}}\right)^{1/\ell} \left(\ell - \frac{n}{p}\right)^{-1/\ell}.
$$

Then it follows from inequality (2.50) that

$$
|u(0)| \leq C |u|_{\ell,p}^{n/\ell p} |u|_{0,p}^{1-n/\ell p}.
$$

This proves inequality (2.47′). □

Proof of Theorem 2.18. By Lemma 2.20, we find that

$$
H^{m,p}(\mathbf{R}^n) \subset B^{m-[n/p]-1}(\mathbf{R}^n),
$$

and

$$
|u|_{j,\infty} \leq C |u|_{m,p}^{(n+jp)/mp} |u|_{0,p}^{1-(n+jp)/mp}, \quad u \in H^{m,p}(\mathbf{R}^n). \tag{2.46}
$$

Hence we have, for $0 \leq k < j$,
(2.51)

$$
\begin{aligned}
|u|_{k,\infty} &\leq C |u|_{m,p}^{(n+kp)/mp} |u|_{0,p}^{1-(n+kp)/mp} \\
&= C \left(|u|_{m,p}^{(n+jp)/mp} |u|_{0,p}^{1-(n+jp)/mp}\right)^{(n+kp)/(n+jp)} |u|_{0,p}^{1-(n+kp)/(n+jp)} \\
&\leq C \left(|u|_{m,p}^{(n+jp)/mp} |u|_{0,p}^{1-(n+jp)/mp} + |u|_{0,p}\right).
\end{aligned}
$$

Therefore, combining inequalities (2.46) and (2.51), we obtain that

$$
\|u\|_{j,\infty} \leq C \left(|u|_{m,p}^{(n+jp)/mp} |u|_{0,p}^{1-(n+jp)/mp} + |u|_{0,p}\right)
$$

$$\leq C\|u\|_{m,p}, \quad 0 \leq j \leq m - [n/p] - 1.$$

This proves inequality (2.45).

On the other hand, inequality (2.44) follows from Theorem 2.15 with

$$q = p, \quad j = 0, \quad a = 1.$$

Indeed, we have

$$|u|_{m-n/p,\infty} = |u|_{-n/r,\infty} = |u|_r \leq C|u|_{m,p}.$$

The proof of Theorem 2.18 is now complete. □

The next imbedding theorem asserts the existence of imbeddings of the spaces $H^{m,p}(\mathbf{R}^n)$ into the spaces $H^{j,r}(\mathbf{R}^n)$, $0 \leq j < m$:

Theorem 2.21. *Let $1 \leq p < \infty$ and let j, m be integers such that $0 \leq j < m$. If $m - j - n/p$ is not a non-negative integer and if*

$$\frac{1}{r} = \frac{j}{n} + \frac{1}{p} - a\frac{m}{n}, \quad \frac{j}{m} \leq a \leq 1,$$

then we have

$$H^{m,p}(\mathbf{R}^n) \subset H^{j,r}(\mathbf{R}^n),$$

with continuous injection.

Moreover, we have

$$\|u\|_{j,r} \leq C\left(|u|^a_{m,p}|u|^{1-a}_{0,p} + |u|_{0,p}\right), \quad u \in H^{m,p}(\mathbf{R}^n). \tag{2.52}$$

Here $C = C(m, n, p, q, j, a)$ is a positive constant.

Proof. We remark that

$$\frac{1}{r} = \frac{1}{p} + \frac{1}{n}(j - am) \geq \frac{1}{p},$$

so that

$$r \geq p \geq 1.$$

If we take

$$q = p$$

in Theorem 2.15, then we have, for $j/m \leq a \leq 1$,

$$|u|_{j,r} \leq C|u|^a_{m,p}|u|^{1-a}_{0,p}, \quad u \in H^{m,p}(\mathbf{R}^n). \tag{2.53}$$

In particular, we have, for $a = j/m$,

$$|u|_{j,p} \leq C|u|^{j/m}_{m,p}|u|^{1-j/m}_{0,p}. \tag{2.54}$$

Now we take an arbitrary integer k such that

$$0 \leq k < j,$$

and let

$$b = \frac{am - j + k}{m - j + k}.$$

Then we have

$$\frac{k}{m - j + k} \leq b \leq 1,$$

and

$$\frac{1}{r} = \frac{k}{n} + \frac{1}{p} - b\frac{m - j + k}{n}.$$

Hence, applying inequality (2.53) with

$$m = m - j + k, \quad j = k, \quad a = b,$$

we obtain that

$$|u|_{k,r} \leq C|u|_{m-j+k,p}^{b}|u|_{0,p}^{1-b}. \tag{2.53'}$$

But, applying inequality (2.54) with

$$j := m - j + k,$$

we obtain that

$$|u|_{m-j+k,p} \leq C|u|_{m,p}^{(m-j+k)/m}|u|_{0,p}^{1-(m-j+k)/m}. \tag{2.54'}$$

Thus, combining inequalities (2.53′) and (2.54′), we find that

(2.55)
$$\begin{aligned}
|u|_{k,r} &\leq C\left(|u|_{m,p}^{(m-j+k)/m}|u|_{0,p}^{(j-k)/m}\right)^{(am-j+k)/(m-j+k)} |u|_{0,p}^{m(1-a)/(m-j+k)} \\
&= C\left(|u|_{m,p}^{a}|u|_{0,p}^{1-a}\right)^{(am-j+k)/am} |u|_{0,p}^{(j-k)/am} \\
&\leq C\left(|u|_{m,p}^{a}|u|_{0,p}^{1-a} + |u|_{0,p}\right), \quad 0 \leq k < j.
\end{aligned}$$

Therefore, inequality (2.52) follows from inequalities (2.55) and (2.53). Indeed, we have

$$\|u\|_{j,r} = \left(\sum_{k=0}^{j} |u|_{k,r}\right)^{1/r} \leq C\left(|u|_{m,p}^{a}|u|_{0,p}^{1-a} + |u|_{0,p}\right).$$

The proof of Theorem 2.21 is complete. □

Corollary 2.22. *Let $1 \leq p < \infty$ and let j, m be integers such that $0 \leq j < m$ and that $m - j - n/p$ is not a non-negative integer. If $p \leq r \leq \infty$ and $j/n + 1/p - m/n \leq 1/r$, then we have*

$$H^{m,p}(\mathbf{R}^n) \subset H^{j,r}(\mathbf{R}^n),$$

with continuous injection.

Proof. We let

$$a = \frac{n}{m}\left(\frac{j}{n} + \frac{1}{p} - \frac{1}{r}\right).$$

Then we find that

$$a = \frac{n}{m}\left(\frac{j}{n} + \frac{1}{p} - \frac{1}{r}\right) \leq \frac{n}{m} \cdot \frac{m}{n} = 1,$$

$$a = \frac{j}{m} + \frac{n}{m}\left(\frac{1}{p} - \frac{1}{r}\right) \geq \frac{j}{m},$$

and that

$$\frac{j}{n} + \frac{1}{p} - a\frac{m}{n} = \frac{j}{n} + \frac{1}{p} - \left(\frac{j}{n} + \frac{1}{p} - \frac{1}{r}\right) = \frac{1}{r}.$$

Hence Corollary 2.22 is an immediate consequence of Theorem 2.21. □

2.4 Imbeddings of the Spaces $H^{m,p}(\Omega)$

The purpose of this section is to prove Sobolev imbedding theorems for a bounded C^∞ domain Ω.

The next theorem, due to Seeley [Se1], asserts that the functions in $C^\infty(\overline{\Omega})$ are the restrictions to Ω of functions in $C^\infty(\mathbf{R}^n)$:

Theorem 2.23. *Let Ω be either the half space $\mathbf{R}^n_+$ or a C^∞ domain in $\mathbf{R}^n$ with bounded boundary Γ. Then there exists a continuous linear extension operator*

$$E : C^\infty(\overline{\Omega}) \longrightarrow C^\infty(\mathbf{R}^n).$$

Furthermore, the restriction to the space $C_0^\infty(\overline{\Omega})$ of E is a continuous linear extension operator on $C_0^\infty(\overline{\Omega})$ into $C_0^\infty(\mathbf{R}^n)$.

Proof. (i) First we consider the case $\Omega = \mathbf{R}^n_+$. The proof is based on the following:

Lemma 2.24. *There exists a function w in the space $\mathcal{S}(\mathbf{R})$ such that*

$$\begin{cases} \operatorname{supp} w \subset [1, \infty), \\ \int_1^\infty t^n\, w(t)\, dt = (-1)^n, \quad n = 0, 1, 2, \cdots. \end{cases}$$

Assuming Lemma 2.24 for the moment, we shall prove Theorem 2.23.

By using the function w in Lemma 2.24, we can define a linear operator

$$E : C^\infty(\overline{\mathbf{R}^n_+}) \longrightarrow C^\infty(\mathbf{R}^n)$$

by the formula

$$E\varphi(x', x_n) = \begin{cases} \varphi(x', x_n) & \text{if } x_n \geq 0, \\ \int_1^\infty w(s)\, \theta(-x_n s)\, \varphi(x', -s x_n)\, ds & \text{if } x_n < 0, \end{cases}$$

where

$$x = (x', x_n) \in \mathbf{R}^n, \quad x' = (x_1, x_2, \cdots, x_{n-1}) \in \mathbf{R}^{n-1},$$

and

$$\begin{cases} \theta \in C_0^\infty(\mathbf{R}), \\ \operatorname{supp} \theta \subset [-2, 2], \\ \theta(t) = 1 \text{ for } |t| \leq 1. \end{cases}$$

Then it is easy to verify the following:

(1) $E\varphi \in C^\infty(\mathbf{R}^n)$.

(2) The operator E maps $C^\infty(\overline{\mathbf{R}^n_+})$ continuously into $C^\infty(\mathbf{R}^n)$.

(3) If $\operatorname{supp} \varphi \subset \{x \in \mathbf{R}^n : |x'| \leq r, 0 \leq x_n \leq a\}$ for some $r > 0$ and $a > 0$, then it follows that $\operatorname{supp} E\varphi \subset \{x \in \mathbf{R}^n : |x'| \leq r, |x_n| \leq a\}$.

This proves Theorem 2.23 for the half space $\mathbf{R}^n_+$.

(ii) Now assume that Ω is a C^∞ domain in $\mathbf{R}^n$ with bounded boundary Γ. Then we can choose a finite covering $\{U_j\}_{j=1}^N$ of Γ by open subsets of $\mathbf{R}^n$ and C^∞ diffeomorphisms χ_j of U_j onto the unit ball $B = \{x \in \mathbf{R}^n : |x| < 1\}$ such that the open sets

$$V_j = \chi_j^{-1}\left(\left\{x \in \mathbf{R}^n : |x'| < \frac{1}{2}, |x_n| < \frac{\sqrt{3}}{2}\right\}\right), \quad 1 \leq j \leq N,$$

form an open covering of the tubular neighborhood

$$\Omega_\delta = \{x \in \Omega : \operatorname{dist}(x, \Gamma) < \delta\}$$

for some $\delta > 0$. Further we can choose an open set V_0 in Ω, bounded away from Γ, such that (see Figure 2.1)

$$\Omega \subset V_0 \bigcup \left(\bigcup_{j=1}^N V_j\right).$$

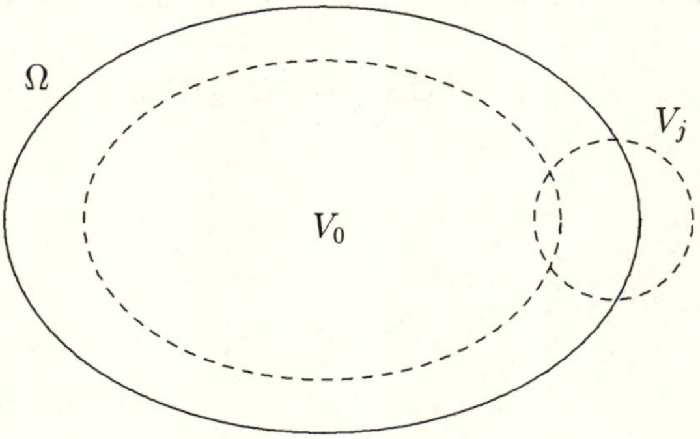

Figure 2.1

Let $\{\omega_j\}_{j=0}^N$ be a partition of unity subordinate to the covering $\{V_j\}_{j=0}^N$. If $\varphi \in C^\infty(\overline{\Omega})$, we define

$$E\varphi = \omega_0\varphi + \sum_{j=1}^{N} \chi_j^* \left(E\left((\chi_j^{-1})^*(\omega_j\varphi)\right)\right).$$

Then it is easy to verify that this operator E enjoys the desired properties.

Theorem 2.23 is proved, apart from the proof of Lemma 2.24. □

Proof of Lemma 2.24. (1) First take a function $u_0 \in C_0^\infty(\mathbf{R})$ such that

$$\begin{cases} u_0 \geq 0 \text{ on } \mathbf{R}, \\ \operatorname{supp} u_0 \subset [1,2], \\ \int_0^\infty u_0(t)\,dt = 1. \end{cases} \tag{2.56}$$

If k is a positive integer, we let

$$u_k(t) = \frac{1}{2^k}\, u_0\left(\frac{t}{2^k}\right). \tag{2.57}$$

Then we have the following:

$$\begin{cases} u_k \geq 0 \text{ on } \mathbf{R}, \\ \operatorname{supp} u_k \subset [2^k, 2^{k+1}], \\ \int_0^\infty u_k(t)\,dt = 1. \end{cases}$$

Hence, for any sequence $\{a_k\}$, the formal sum

$$w(t) = \sum_{k=0}^{\infty} a_k\, u_k(t)$$

make a sense. We shall choose a sequence $\{a_k\}$ in such a way that the function w has the desired properties.

(2) Next fix an integer $N > 0$ and construct the N-th approximation to the function w:

$$w_N(t) = \sum_{k=0}^{N} a_{Nk}\, u_k(t).$$

The coefficients a_{Nk} will be picked to satisfy the conditions

$$\int_0^\infty t^n\, w_N(t)\, dt = (-1)^n, \quad n = 0, 1, \cdots, N. \tag{2.58}$$

But, by conditions (2.56) and (2.57), this equation may be rewritten as

$$\sum_{k=0}^{N} 2^{nk}\, a_{Nk} = h_n, \quad n = 0, 1, \cdots, N, \tag{2.58$'$}$$

where

$$h_n = \frac{(-1)^n}{\int_0^\infty t^n\, u_0(t)\, dt}.$$

Now equation (2.58′) is a linear system of $N + 1$ equations in $N + 1$ unknowns, and its determinant is the Vandermonde determinant:

$$\Delta_N = \begin{vmatrix} 1 & 1 & \cdots & 1 \\ 1 & 2 & \cdots & 2^N \\ \vdots & \vdots & \ddots & \vdots \\ 1 & 2^N & \cdots & 2^{N^2} \end{vmatrix} = \prod_{\substack{i,j=0 \\ i<j}}^{N} (2^j - 2^i).$$

Hence, using Cramer's rule, we find the solutions $\{a_{Nk}\}$ of equation (2.58′) as follows.

$$a_{Nk} = (-1)^k \Delta_{Nk} \left(\sum_{\ell=0}^{N} (-1)^\ell\, h_\ell\, S_{Nk}^{N-\ell} \right) \Delta^{-1},$$

where

$$\Delta_{Nk} = \prod_{\substack{i,j=0 \\ i,j \neq k \\ i<j}}^{N} (2^j - 2^i),$$

and

$$S_{Nk}^m = \text{the elementary symmetric polynomial of degree } m$$
$$\text{in the elements } \{1, 2, \cdots, 2^{k-1}, 2^{k+1}, \cdots, 2^N\}.$$

Since we have by condition (2.56)

$$|h_\ell| \le 1$$

and

$$\sum_{\ell=0}^{N} S_{Nk}^{N-\ell} = \prod_{\substack{\ell=0 \\ \ell \neq k}}^{N} (1+2^\ell),$$

it follows that

$$|a_{Nk}| \le \left\{ \prod_{\ell=0}^{k-1} (1+2^\ell) \prod_{\ell=k+1}^{N} (1+2^\ell) \right\} \left\{ \prod_{\ell=0}^{k-1} (2^k - 2^\ell) \prod_{\ell=k+1}^{N} (2^\ell - 2^k) \right\}^{-1}$$
$$=: A_k \cdot B_{Nk}$$

where

$$A_k = \prod_{\ell=0}^{k-1} \frac{1+2^\ell}{2^\ell - 2^k},$$
$$B_{Nk} = \prod_{\ell=k+1}^{N} \frac{1+2^\ell}{2^\ell - 2^k}.$$

But we remark that

(2.59) $$|A_k| \le \prod_{\ell=0}^{k-1} \frac{2^{\ell+1}}{2^{k-1}} = 2^{(3k-k^2)/2}$$

and also

(2.60) $$\begin{aligned} \log B_{Nk} &= \sum_{\ell=k+1}^{N} \log \left(1 + \frac{1+2^k}{2^\ell - 2^k} \right) \\ &< \sum_{\ell=k+1}^{N} \left(1 + \frac{1+2^k}{2^\ell - 2^k} \right) \\ &< \left(1+2^k\right) \sum_{\ell=k+1}^{N} \frac{1}{2^{\ell-1}} \\ &< 4, \end{aligned}$$

since we have the inequality $\log(1+x) < x$ for all $x > 0$.

Therefore, by combining estimates (2.59) and (2.60), we obtain that

(2.61) $$|a_{Nk}| \le e^4\, 2^{(3k-k^2)/2}, \quad k = 0, 1, \cdots, N.$$

To prove that a finite $\lim_{N\to\infty} a_{Nk}$ exists for every integer $k \geq 0$, it suffices to show that each sequence $\{a_{Nk}\}_{N=1}^{\infty}$ is a Cauchy sequence.

Since we have

$$S_{N+1\,k}^{N+1-\ell} = 2^{N+1}\, S_{Nk}^{N-\ell} + S_{Nk}^{N+1-\ell},$$

it follows that

$$a_{N+1\,k} - a_{Nk} = (-1)^k \left\{ \sum_{\ell=0}^{N} \left[(-1)^\ell\, 2^k\, h_\ell + (-1)^{\ell+1}\, h_{\ell+1} \right] S_{Nk}^{N-\ell} \right\} \times \Delta_{N+1\,k} \cdot \Delta_{N+1}^{-1},$$

so that

$$|a_{N+1\,k} - a_{Nk}| \leq \left(1 + 2^k\right) \left(\sum_{\ell=0}^{N} S_{Nk}^{N-\ell} \right) \Delta_{N+1\,k} \cdot \Delta_{N+1}^{-1} = |A_k| \cdot B_{Nk} \left(2^{N+1} - 2^k\right)^{-1}.$$

Hence we obtain from estimates (2.59) and (2.60) that for any integer $M > 0$

$$|a_{N+M\,k} - a_{Nk}| \leq \left(1 + 2^k\right) |A_k|\, e^4 \sum_{m=1}^{M} \left(2^{N+m} - 2^k\right)^{-1} \leq e^4 \left(1 + 2^k\right) |A_k| \cdot 2^{-N}.$$

This proves that the sequence $\{a_{Nk}\}_{N=1}^{\infty}$ is a Cauchy sequence for every integer $k \geq 0$.

(3) Finally we let

$$a_k = \lim_{N\to\infty} a_{Nk}, \quad k = 0, 1, \cdots.$$

Then, letting $N \to \infty$ in estimate (2.61), we have

$$|a_k| \leq e^4\, 2^{(3k-k^2)/2}. \tag{2.62}$$

In view of assertions (2.61), (2.62) and (2.58), it is easy to verify that $w_N \to w$ in the space $\mathcal{S}(\mathbf{R})$ and that the limit function w enjoys the desired properties.

Now the proof of Lemma 2.24, and hence that of Theorem 2.23, is complete. □

By using Theorem 2.23, we can construct an extension operator

$$E : H^{m,p}(\Omega) \longrightarrow H^{m,p}(\mathbf{R}^n).$$

Thus it is easy to see that

$H^{m,p}(\Omega) =$ the space of restrictions to Ω of functions in $H^{m,p}(\mathbf{R}^n)$

with the norm

$$\|u\|_{m,p,\Omega} = \inf \|U\|_{m,p,\mathbf{R}^n},$$

where the infimum is taken over all $U \in H^{m,p}(\mathbf{R}^n)$ which equal u in Ω. Furthermore, we have for all functions $u \in H^{m,p}(\Omega)$

$$\|u\|_{j,r,\Omega} \leq \|Eu\|_{j,r,\mathbf{R}^n} \leq C\|Eu\|_{m,p,\mathbf{R}^n} \leq CC'\|u\|_{m,p,\Omega}.$$

Therefore, by such extension arguments, one can prove the following:

Theorem 2.25. *Theorems 2.15, 2.18, 2.21 and Corollary 2.22 remain valid for a bounded* C^∞ *domain* Ω.

CHAPTER III

L^p THEORY OF PSEUDO-DIFFERENTIAL OPERATORS

In this chapter we present a brief description of the basic concepts and results of the L^p theory of pseudo-differential operators – a modern theory of potentials – which will be used in subsequent chapters. For detailed studies of pseudo-differential operators, the reader is referred to Chazarain and Piriou [CP], Kumano-go [Ku] and Taylor [Ty].

3.1 Generalized Sobolev Spaces and Besov Spaces

Let Ω be a bounded domain of Euclidean space $\mathbf{R}^n$ with C^∞ boundary Γ. Its closure $\overline{\Omega} = \Omega \cup \Gamma$ is an n-dimensional compact C^∞ manifold with boundary. One may assume that (see Figure 3.1):

(a) The domain Ω is a relatively compact open subset of an n-dimensional compact C^∞ manifold M without boundary.

(b) In a neighborhood W of Γ in M a normal coordinate t is chosen so that the points of W are represented as (x', t), $x' \in \Gamma$, $-1 < t < 1$; $t > 0$ in Ω, $t < 0$ in $M \setminus \overline{\Omega}$ and $t = 0$ only on Γ.

(c) The manifold M is equipped with a strictly positive density μ which, on W, is the product of a strictly positive density ω on Γ and the Lebesgue measure dt on $(-1, 1)$. This manifold M is called the *double* of Ω.

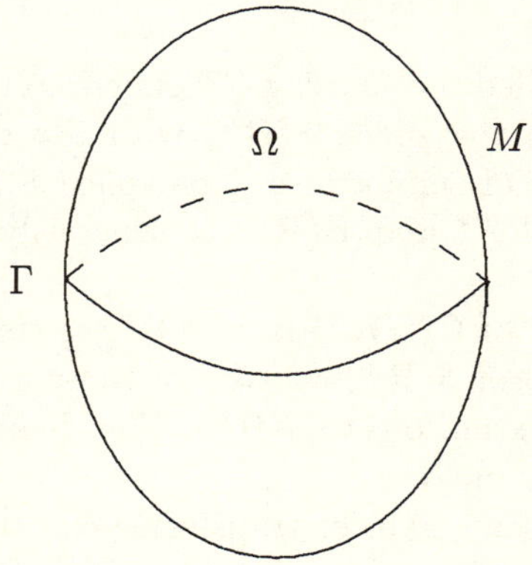

Figure 3.1

The function spaces we shall treat are the following (cf. [BL], [Tr]):

Typeset by $\mathcal{A}_\mathcal{M}\mathcal{S}$-TEX

(i) The generalized Sobolev spaces $H^{s,p}(\Omega)$ and $H^{s,p}(M)$, consisting of all potentials of order s of L^p functions. When s is integral, these spaces coincide with the usual Sobolev spaces $W^{s,p}(\Omega)$ and $W^{s,p}(M)$.

(ii) The Besov spaces $B^{s,p}(\Gamma)$. These are function spaces defined in terms of the L^p modulus of continuity, and enter naturally in connection with boundary value problems.

First we recall the basic definitions and facts about the Fourier transform.

If $f \in L^1(\mathbf{R}^n)$, we define its (direct) Fourier transform $\mathcal{F}f$ by the formula

$$\mathcal{F}f(\xi) = \int_{\mathbf{R}^n} e^{-ix\cdot\xi} f(x)dx, \quad \xi = (\xi_1, \xi_2, \cdots, \xi_n),$$

where $x \cdot \xi = x_1\xi_1 + \cdots + x_n\xi_n$. We also denote $\mathcal{F}f$ by $\hat{f}$.

Similarly, if $g \in L^1(\mathbf{R}^n)$, we define

$$\mathcal{F}^*g(x) = \frac{1}{(2\pi)^n} \int_{\mathbf{R}^n} e^{ix\cdot\xi} g(\xi)d\xi.$$

The function $\mathcal{F}^*g$ is called the inverse Fourier transform of g.

We introduce a subspace of $L^1(\mathbf{R}^n)$ which is invariant under the Fourier transform.

We let

$\mathcal{S}(\mathbf{R}^n)$ = the space of C^∞ functions φ on $\mathbf{R}^n$ such that we have for any non-negative integer j

$$p_j(\varphi) = \sup_{\substack{x\in\mathbf{R}^n \\ |\alpha|\leq j}} \left\{(1 + |x|^2)^{j/2} |\partial^\alpha \varphi(x)|\right\} < \infty.$$

The space $\mathcal{S}(\mathbf{R}^n)$ is called the space of C^∞ functions on $\mathbf{R}^n$ *rapidly decreasing at infinity.* We equip the space $\mathcal{S}(\mathbf{R}^n)$ with the topology defined by the countable family $\{p_j\}$ of seminorms. The space $\mathcal{S}(\mathbf{R}^n)$ is a Fréchet space. The transforms $\mathcal{F}$ and $\mathcal{F}^*$ map $\mathcal{S}(\mathbf{R}^n)$ continuously into itself, and $\mathcal{F}\mathcal{F}^* = \mathcal{F}^*\mathcal{F} = I$ on $\mathcal{S}(\mathbf{R}^n)$.

Since the injection of $C_0^\infty(\mathbf{R}^n)$ into $\mathcal{S}(\mathbf{R}^n)$ is continuous, it follows that the dual space $\mathcal{S}'(\mathbf{R}^n)$ of $\mathcal{S}(\mathbf{R}^n)$ consists of those distributions $T \in \mathcal{D}'(\mathbf{R}^n)$ that have continuous extensions to $\mathcal{S}(\mathbf{R}^n)$. The elements of $\mathcal{S}'(\mathbf{R}^n)$ are called tempered distributions on $\mathbf{R}^n$.

The direct and inverse Fourier transforms can be extended to the space $\mathcal{S}'(\mathbf{R}^n)$ by the following formulas:

$$\langle \mathcal{F}u, \varphi\rangle = \langle u, \mathcal{F}\varphi\rangle, \quad \varphi \in \mathcal{S}(\mathbf{R}^n);$$
$$\langle \mathcal{F}^*u, \varphi\rangle = \langle u, \mathcal{F}^*\varphi\rangle, \quad \varphi \in \mathcal{S}(\mathbf{R}^n).$$

Once again, the transforms $\mathcal{F}$ and $\mathcal{F}^*$ map $\mathcal{S}'(\mathbf{R}^n)$ continuously into itself, and $\mathcal{F}\mathcal{F}^* = \mathcal{F}^*\mathcal{F} = I$ on $\mathcal{S}'(\mathbf{R}^n)$.

If $s \in \mathbf{R}$, we define a linear map

$$J^s : \mathcal{S}'(\mathbf{R}^n) \longrightarrow \mathcal{S}'(\mathbf{R}^n)$$

by the formula

$$J^s u = \mathcal{F}^* \left((1 + |\xi|^2)^{-s/2} \mathcal{F} u \right), \quad u \in \mathcal{S}'(\mathbf{R}^n).$$

Then the map J^s is an isomorphism of $\mathcal{S}'(\mathbf{R}^n)$ onto itself, and its inverse is the map J^{-s}. The function $J^s u$ is called the Bessel potential of order s of u.

Now, if $s \in \mathbf{R}$ and $1 < p < \infty$, we let

$$H^{s,p}(\mathbf{R}^n) = \text{the image of } L^p(\mathbf{R}^n) \text{ under the mapping } J^s.$$

We equip $H^{s,p}(\mathbf{R}^n)$ with the norm $\|u\|_{s,p} = \|J^{-s}u\|_p$ for $u \in H^{s,p}(\mathbf{R}^n)$. The space $H^{s,p}(\mathbf{R}^n)$ is called the (generalized) *Sobolev space* of order s.

We list some basic topological properties of $H^{s,p}(\mathbf{R}^n)$:

(1) The space $\mathcal{S}(\mathbf{R}^n)$ is dense in $H^{s,p}(\mathbf{R}^n)$.

(2) The space $H^{-s,p'}(\mathbf{R}^n)$ is the dual space of $H^{s,p}(\mathbf{R}^n)$, where $p' = p/(p-1)$ is the exponent conjugate to p.

(3) If $s > t$, then we have the inclusions

$$\mathcal{S}(\mathbf{R}^n) \subset H^{s,p}(\mathbf{R}^n) \subset H^{t,p}(\mathbf{R}^n) \subset \mathcal{S}'(\mathbf{R}^n),$$

with continuous injections.

(4) If s is a non-negative integer, then the space $H^{s,p}(\mathbf{R}^n)$ is isomorphic to the usual Sobolev space $W^{s,p}(\mathbf{R}^n)$, that is, the space $H^{s,p}(\mathbf{R}^n)$ coincides with the space of functions $u \in L^p(\mathbf{R}^n)$ such that $D^\alpha u \in L^p(\mathbf{R}^n)$ for $|\alpha| \leq s$, and the norm $\|\cdot\|_{s,p}$ is equivalent to the norm

$$\left(\sum_{|\alpha| \leq s} \int_{\mathbf{R}^n} |D^\alpha u(x)|^p dx \right)^{1/p}.$$

Next, if $1 < p < \infty$, we let

$$\begin{aligned} B^{1,p}(\mathbf{R}^{n-1}) = {} & \text{the space of (equivalence classes of) functions} \\ & \varphi \in L^p(\mathbf{R}^{n-1}) \text{ for which} \\ & \iint_{\mathbf{R}^{n-1} \times \mathbf{R}^{n-1}} \frac{|\varphi(x+y) - 2\varphi(x) + \varphi(x-y)|^p}{|y|^{n-1+p}} \, dy \, dx < \infty. \end{aligned}$$

The space $B^{1,p}(\mathbf{R}^{n-1})$ is a Banach space with respect to the norm

$$|\varphi|_{1,p} = \left(\int_{\mathbf{R}^{n-1}} |\varphi(x)|^p dx \right.$$

$$+ \iint_{\mathbf{R}^{n-1}\times\mathbf{R}^{n-1}} \frac{|\varphi(x+y) - 2\varphi(x) + \varphi(x-y)|^p}{|y|^{n-1+p}}\, dy\, dx\Big)^{1/p}.$$

If $s \in \mathbf{R}$, we let

$$B^{s,p}(\mathbf{R}^{n-1}) = \text{the image of } B^{1,p}(\mathbf{R}^{n-1}) \text{ under the mapping } {J'}^{s-1}, \text{ where } {J'}^{s-1} \text{ is the Bessel potential of order } s-1 \text{ on } \mathbf{R}^{n-1}.$$

We equip the space $B^{s,p}(\mathbf{R}^{n-1})$ with the norm $|\varphi|_{s,p} = \left|{J'}^{-s+1}\varphi\right|_{1,p}$ for $\varphi \in B^{s,p}(\mathbf{R}^{n-1})$. The space $B^{s,p}(\mathbf{R}^{n-1})$ is called the *Besov space* of order s.

We list some basic topological properties of $B^{s,p}(\mathbf{R}^{n-1})$:

(1) The space $\mathcal{S}(\mathbf{R}^{n-1})$ is dense in $B^{s,p}(\mathbf{R}^{n-1})$.

(2) The space $B^{-s,p'}(\mathbf{R}^{n-1})$ is the dual space of $B^{s,p}(\mathbf{R}^{n-1})$, where $p' = p/(p-1)$.

(3) If $s > t$, then we have the inclusions

$$\mathcal{S}(\mathbf{R}^{n-1}) \subset B^{s,p}(\mathbf{R}^{n-1}) \subset B^{t,p}(\mathbf{R}^{n-1}) \subset \mathcal{S}'(\mathbf{R}^{n-1}),$$

with continuous injections.

(4) If $s = m + \sigma$ where m is a non-negative integer and $0 < \sigma < 1$, then the Besov space $B^{s,p}(\mathbf{R}^{n-1})$ coincides with the space of functions $\varphi \in H^{m,p}(\mathbf{R}^{n-1})$ such that for $|\alpha| = m$

$$\iint_{\mathbf{R}^{n-1}\times\mathbf{R}^{n-1}} \frac{|D^\alpha\varphi(x) - D^\alpha\varphi(y)|^p}{|x-y|^{n-1+p\sigma}}\, dx\, dy < \infty.$$

Furthermore, the norm $|\varphi|_{s,p}$ is equivalent to the norm

$$\Big(\sum_{|\alpha|\le m} \int_{\mathbf{R}^{n-1}} |D^\alpha\varphi(x)|^p dx + \sum_{|\alpha|=m} \iint_{\mathbf{R}^{n-1}\times\mathbf{R}^{n-1}} \frac{|D^\alpha\varphi(x) - D^\alpha\varphi(y)|^p}{|x-y|^{n-1+p\sigma}}\, dx\, dy\Big)^{1/p}.$$

Now we define the generalized Sobolev spaces $H^{s,p}(\Omega)$, $H^{s,p}(M)$ and the Besov spaces $B^{s,p}(\Gamma)$ for arbitrary values of s.

For each $s \in \mathbf{R}$, we define

$$H^{s,p}(\Omega) = \text{the space of restrictions to } \Omega \text{ of functions in } H^{s,p}(\mathbf{R}^n).$$

We equip the space $H^{s,p}(\Omega)$ with the norm

$$\|u\|_{s,p} = \inf \|U\|_{s,p},$$

where the infimum is taken over all $U \in H^{s,p}(\mathbf{R}^n)$ which equal u in Ω. The space $H^{s,p}(\Omega)$ is a Banach space with respect to the norm $\|\cdot\|_{s,p}$. We remark that

$$H^{0,p}(\Omega) = L^p(\Omega), \quad \|\cdot\|_{0,p} = \|\cdot\|_p.$$

The spaces $H^{s,p}(M)$ are defined to be locally the spaces $H^{s,p}(\mathbf{R}^n)$, upon using local coordinate systems flattening out M, together with a partition of unity. The spaces $B^{s,p}(\Gamma)$ are defined similarly, with $H^{s,p}(\mathbf{R}^n)$ replaced by $B^{s,p}(\mathbf{R}^{n-1})$. The norms of $H^{s,p}(M)$ and $B^{s,p}(\Gamma)$ will be denoted respectively by $\|\cdot\|_{s,p}$ and $|\cdot|_{s,p}$.

We state two important facts which will be used in the study of boundary value problems:

(I) The restriction map

$$\begin{aligned} \rho : H^{s,p}(\Omega) &\longrightarrow B^{s-1/p,p}(\Gamma) \\ u &\longmapsto u|_\Gamma \end{aligned}$$

is well defined for all $s > 1/p$, and is surjective (cf. [St]).

(II) (Rellich) If $s > t$, then the injections

$$H^{s,p}(M) \longrightarrow H^{t,p}(M),$$
$$B^{s,p}(\Gamma) \longrightarrow B^{t,p}(\Gamma)$$

are both compact (or completely continuous).

Finally we introduce a space of distributions on Ω which behave locally just like the distributions in $H^{s,p}(\mathbf{R}^n)$:

$$\begin{aligned} H^{s,p}_{loc}(\Omega) = &\text{ the space of distributions } u \in \mathcal{D}'(\Omega) \text{ such that} \\ &\varphi u \in H^{s,p}(\mathbf{R}^n) \text{ for all } \varphi \in C_0^\infty(\Omega). \end{aligned}$$

We equip the space $H^{s,p}_{loc}(\Omega)$ with the topology defined by the seminorms $u \mapsto \|\varphi u\|_{s,p}$ as φ ranges over $C_0^\infty(\Omega)$. It is easy to verify that $H^{s,p}_{loc}(\Omega)$ is a Fréchet space.

3.2 Fourier Integral Operators

In this section, we present a brief description of the basic concepts and results of the theory of Fourier integral operators.

3.2A Symbol Classes. Let Ω be an open subset of $\mathbf{R}^n$. If $m \in \mathbf{R}$ and $0 \le \delta < \rho \le 1$, we let

$$\begin{aligned} S^m_{\rho,\delta}(\Omega \times \mathbf{R}^N) = &\text{ the set of all functions } a \in C^\infty(\Omega \times \mathbf{R}^N) \text{ with the} \\ &\text{property that, for any compact } K \subset \Omega \text{ and any} \end{aligned}$$

multi-indices α, β, there exists a constant $C_{K,\alpha,\beta} > 0$ such that we have for all $x \in K$ and $\theta \in \mathbf{R}^N$

$$\left|\partial_\theta^\alpha \partial_x^\beta a(x,\theta)\right| \le C_{K,\alpha,\beta}(1+|\theta|)^{m-\rho|\alpha|+\delta|\beta|}.$$

The elements of $S^m_{\rho,\delta}(\Omega \times \mathbf{R}^N)$ are called *symbols* of order m. We drop the $\Omega \times \mathbf{R}^N$ and use $S^m_{\rho,\delta}$ when the context is clear.

Examples 3.1. (1) A polynomial $p(x,\xi) = \sum_{|\alpha|\le m} a_\alpha(x)\xi^\alpha$ of order m with coefficients in $C^\infty(\Omega)$ is in $S^m_{1,0}(\Omega \times \mathbf{R}^n)$.

(2) If $m \in \mathbf{R}$, the function

$$\Omega \times \mathbf{R}^n \ni (x,\xi) \longmapsto \left(1+|\xi|^2\right)^{m/2}$$

is in $S^m_{1,0}(\Omega \times \mathbf{R}^n)$.

(3) A function $a \in C^\infty(\Omega \times (\mathbf{R}^N \setminus \{0\})$ is said to be positively homogeneous of degree m in θ if it satisfies

$$a(x,t\theta) = t^m\, a(x,\theta), \quad t > 0.$$

If $a(x,\theta)$ is positively homogeneous of degree m in θ and if $\varphi(\theta)$ is a C^∞ function such that $\varphi(\theta) = 0$ for $|\theta| \le 1/2$ and $\varphi(\theta) = 1$ for $|\theta| \ge 1$, then the function $\varphi(\theta)a(x,\theta)$ is in $S^m_{1,0}(\Omega \times \mathbf{R}^N)$.

If K is a compact subset of Ω and j is a non-negative integer, we define a seminorm $p_{K,j,m}$ on $S^m_{\rho,\delta}(\Omega \times \mathbf{R}^N)$ by

$$S^m_{\rho,\delta}(\Omega \times \mathbf{R}^N) \ni a \longmapsto p_{K,j,m}(a) = \sup_{\substack{x\in K \\ \theta\in\mathbf{R}^N \\ |\alpha|\le j}} \frac{\left|\partial_\theta^\alpha \partial_x^\beta a(x,\theta)\right|}{(1+|\theta|)^{m-\rho|\alpha|+\delta|\beta|}}.$$

We equip the space $S^m_{\rho,\delta}(\Omega \times \mathbf{R}^N)$ with the topology defined by the family $\{p_{K,j,m}\}$ of seminorms where K ranges over all compact subsets of Ω and $j = 0, 1, \cdots$. The space $S^m_{\rho,\delta}(\Omega \times \mathbf{R}^N)$ is a Fréchet space.

We set

$$S^{-\infty}(\Omega \times \mathbf{R}^N) = \bigcap_{m\in\mathbf{R}} S^m_{\rho,\delta}(\Omega \times \mathbf{R}^N).$$

The next theorem gives a meaning to a formal sum of symbols of decreasing order:

Theorem 3.2. *Let $a_j \in S^{m_j}_{\rho,\delta}(\Omega \times \mathbf{R}^N)$, $m_j \downarrow -\infty$, $j = 0, 1, \cdots$. Then there exists a symbol $a \in S^{m_0}_{\rho,\delta}(\Omega \times \mathbf{R}^N)$, unique modulo $S^{-\infty}(\Omega \times \mathbf{R}^N)$, such that we have for all $k > 0$*

$$a - \sum_{j=0}^{k-1} a_j \in S^{m_k}_{\rho,\delta}(\Omega \times \mathbf{R}^N). \tag{3.1}$$

If formula (3.1) holds, we write

$$a \sim \sum_{j=0}^{\infty} a_j.$$

The formal sum $\sum_j a_j$ is called an asymptotic expansion of a.

A symbol $a(x,\theta) \in S^m_{1,0}(\Omega \times \mathbf{R}^N)$ is said to be *classical* if there exist C^∞ functions $a_j(x,\theta)$, positively homogeneous of degree $m-j$ in θ for $|\theta| \geq 1$, such that

$$a \sim \sum_{j=0}^{\infty} a_j.$$

The homogeneous function a_0 of degree m is called the principal part of a.

We let

$$S^m_{cl}(\Omega \times \mathbf{R}^N) = \text{ the set of all classical symbols of order } m.$$

For example, the symbols in Examples 3.1 are all classical, and they have respectively as principal part the following functions:

(1) $p_m(x,\xi) = \sum_{|\alpha|=m} a_\alpha(x)\xi^\alpha$.
(2) $|\xi|^m$.
(3) $a(x,\theta)$.

A symbol $a(x,\theta)$ in $S^m_{\rho,\delta}(\Omega \times \mathbf{R}^N)$ is said to be *elliptic* of order m if, for any compact $K \subset \Omega$, there exists a constant $C_K > 0$ such that

$$|a(x,\theta)| \geq C_K(1+|\theta|)^m, \quad x \in K, \ |\theta| \geq \frac{1}{C_K}.$$

There is a simple criterion in the case of classical symbols.

Theorem 3.3. *Let $a(x,\theta)$ be in $S^m_{cl}(\Omega \times \mathbf{R}^N)$ with principal part $a_0(x,\theta)$. Then $a(x,\theta)$ is elliptic if and only if we have*

$$a_0(x,\theta) \neq 0, \ x \in \Omega, \ |\theta| = 1.$$

3.2B Phase Functions. Let Ω be an open subset of $\mathbf{R}^n$. A function $\varphi(x,\theta)$ in $C^\infty\left(\Omega \times \left(\mathbf{R}^N \setminus \{0\}\right)\right)$ is called a *phase function* on $\Omega \times (\mathbf{R}^N \setminus \{0\})$ if it satisfies the following three conditions:

(a) φ is real-valued.
(b) φ is positively homogeneous of degree 1 in the variable θ.
(c) The differential $d\varphi$ does not vanish on $\Omega \times \left(\mathbf{R}^N \setminus \{0\}\right)$.

Example 3.4. Let U be an open subset of $\mathbf{R}^p$ and $\Omega = U \times U$. The function

$$\varphi(x,y,\xi) = (x-y)\cdot\xi$$

is a phase function on the space $\Omega \times (\mathbf{R}^p \setminus \{0\})$ $(n = 2p, \ N = p)$.

The next lemma will play a fundamental role in defining oscillatory integrals.

Lemma 3.5. *If φ is a phase function on $\Omega \times (\mathbf{R}^N \setminus \{0\})$, then there exists a first-order differential operator*

$$L = \sum_{j=1}^{N} a_j(x,\theta)\frac{\partial}{\partial \theta_j} + \sum_{k=1}^{n} b_k(x,\theta)\frac{\partial}{\partial x_k} + c(x,\theta)$$

such that

$$L(e^{i\varphi}) = e^{i\varphi},$$

and its coefficients a_j, b_k, c enjoy the following properties:

$$a_j \in S^0_{1,0}\,;\ b_k,\ c \in S^{-1}_{1,0}.$$

Furthermore, the transpose L' of L has coefficients a'_j, b'_k, c' in the same symbol classes as a_j, b_k, c, respectively.

For example, if φ is a phase function as in Example 3.4,

$$\varphi(x,y,\xi) = (x-y)\cdot\xi, \quad (x,y) \in U \times U,\ \xi \in (\mathbf{R}^p \setminus \{0\}),$$

then the operator L is given by the following:

$$L = \frac{1}{i}\frac{1-\rho(\xi)}{2+|x-y|^2}\left\{\sum_{j=1}^{p}(x_j-y_j)\frac{\partial}{\partial \xi_j} + \sum_{k=1}^{p}\frac{\xi_j}{|\xi|^2}\frac{\partial}{\partial x_k} + \sum_{k=1}^{p}\frac{-\xi_j}{|\xi|^2}\frac{\partial}{\partial y_k}\right\}$$
$$+ \rho(\xi),$$

where $\rho(\xi)$ is a function in $C_0^\infty(\mathbf{R}^p)$ such that $\rho = 1$ for $|\xi| \leq 1$.

3.2C Oscillatory Integrals. We let

$$S^\infty_{\rho,\delta}\left(\Omega \times \mathbf{R}^N\right) = \bigcup_{m \in \mathbf{R}} S^m_{\rho,\delta}\left(\Omega \times \mathbf{R}^N\right).$$

If $\varphi(x,\theta)$ is a phase function on $\Omega \times (\mathbf{R}^N \setminus \{0\})$, we wish to give a meaning to the integral

$$(3.2) \qquad I_\varphi(au) = \iint_{\Omega \times \mathbf{R}^N} e^{i\varphi(x,\theta)} a(x,\theta)u(x)\,dx d\theta, \quad u \in C_0^\infty(\Omega),$$

for each symbol $a(x,\theta) \in S^\infty_{\rho,\delta}\left(\Omega \times \mathbf{R}^N\right)$.

By Lemma 3.5, we can replace $e^{i\varphi}$ in formula (3.2) by $L(e^{i\varphi})$. Then a *formal* integration by parts gives us that

$$I_\varphi(au) = \iint_{\Omega \times \mathbf{R}^N} e^{i\varphi(x,\theta)}\, L'(a(x,\theta)u(x))\,dx\,d\theta.$$

But the properties of the coefficients of L' imply that L' maps $S^r_{\rho,\delta}$ continuously into $S^{r-\eta}_{\rho,\delta}$ for all $r \in \mathbf{R}$, where $\eta = \min(\rho, 1-\delta)$. Continuing this process, one can reduce the growth of the integrand at infinity until it becomes integrable, and give a meaning to the integral (3.2) for each symbol $a \in S^\infty_{\rho,\delta}\left(\Omega \times \mathbf{R}^N\right)$.

More precisely, we have the following:

Theorem 3.6. *(i) The linear functional*

$$S^{-\infty}\left(\Omega \times \mathbf{R}^N\right) \ni a \longmapsto I_\varphi(au) \in \mathbf{C}$$

extends uniquely to a linear functional ℓ on the space $S^{\infty}_{\rho,\delta}\left(\Omega \times \mathbf{R}^N\right)$ whose restriction to each space $S^{m}_{\rho,\delta}\left(\Omega \times \mathbf{R}^N\right)$ is continuous. Further, the restriction to $S^{m}_{\rho,\delta}\left(\Omega \times \mathbf{R}^N\right)$ of ℓ is expressed as

$$\ell(a) = \iint_{\Omega \times \mathbf{R}^N} e^{i\varphi(x,\theta)} \left(L'\right)^k (a(x,\theta)u(x))\, dx\, d\theta,$$

where $k > (m+N)/\eta$, $\eta = \min(\rho, 1-\delta)$.

(ii) For any fixed $a \in S^{m}_{\rho,\delta}\left(\Omega \times \mathbf{R}^N\right)$, the mapping

$$C_0^\infty(\Omega) \ni u \longmapsto I_\varphi(au) = \ell(a) \in \mathbf{C} \tag{3.3}$$

is a distribution of order $\leq k$ for $k > (m+N)/\eta$.

We call the linear functional ℓ on $S^{\infty}_{\rho,\delta}$ an *oscillatory integral*, but use the standard notation as in formula (3.2). The distribution (3.3) is called the *Fourier integral distribution* associated with the phase function φ and the amplitude a, and is denoted by

$$\int_{\mathbf{R}^N} e^{i\varphi(x,\theta)} a(x,\theta)\, d\theta.$$

If u is a distribution on Ω, the *singular support* of u is the smallest closed subset of Ω outside of which u is of class C^∞. The singular support of u is denoted by $\operatorname{sing\,supp} u$.

The next theorem estimates the singular support of a Fourier integral distribution.

Theorem 3.7. *If φ is a phase function on the space $\Omega \times \left(\mathbf{R}^N \setminus \{0\}\right)$ and if a is in $S^{\infty}_{\rho,\delta}\left(\Omega \times \mathbf{R}^N\right)$, then the distribution*

$$A = \int_{\mathbf{R}^N} e^{i\varphi(x,\theta)} a(x,\theta)\, d\theta \in \mathcal{D}'(\Omega)$$

satisfies

$$\operatorname{sing\,supp} A \subset \left\{x \in \Omega : d_\theta\varphi(x,\theta) = 0 \text{ for some } \theta \in \mathbf{R}^N \setminus \{0\}\right\}.$$

3.2D Fourier Integral Operators. Let U and V be open subsets of $\mathbf{R}^p$ and $\mathbf{R}^q$, respectively. If $\varphi(x,y,\theta)$ is a phase function on $U\times V\times(\mathbf{R}^N\setminus\{0\})$ and if $a(x,y,\theta)\in S^{\infty}_{\rho,\delta}(U\times V\times\mathbf{R}^N)$, then there is associated a distribution $K\in\mathcal{D}'(U\times V)$ defined by the formula

$$K=\int_{\mathbf{R}^N} e^{i\varphi(x,y,\theta)}a(x,y,\theta)\,d\theta.$$

Applying Theorem 3.7 to our situation, we obtain that

$$\text{sing supp}\, K\subset\left\{(x,y)\in U\times V : d_\theta\varphi(x,y,\theta)=0 \text{ for some } \theta\in\mathbf{R}^N\setminus\{0\}\right\}.$$

The distribution K defines a continuous linear operator

$$A: C_0^\infty(V)\longrightarrow\mathcal{D}'(U)$$

by the formula

$$\langle Av,u\rangle = \langle K,u\otimes v\rangle,\quad u\in C_0^\infty(U),\ v\in C_0^\infty(V).$$

The operator A is called the *Fourier integral operator* associated with the phase function φ and the amplitude a, and is denoted by

$$Av(x)=\iint_{V\times\mathbf{R}^N} e^{i\varphi(x,y,\theta)}a(x,y,\theta)v(y)\,dy\,d\theta,\quad v\in C_0^\infty(V).$$

The next theorem summarizes some basic properties of the operator A.

Theorem 3.8. *(i) If $d_{y,\theta}\varphi(x,y,\theta)\neq 0$ on $U\times V\times\left(\mathbf{R}^N\setminus\{0\}\right)$, then the operator A maps $C_0^\infty(V)$ continuously into $C^\infty(U)$.*

(ii) If $d_{x,\theta}\varphi(x,y,\theta)\neq 0$ on $U\times V\times\left(\mathbf{R}^N\setminus\{0\}\right)$, then the operator A extends to a continuous linear operator on $\mathcal{E}'(V)$ into $\mathcal{D}'(U)$.

(iii) If $d_{y,\theta}\varphi(x,y,\theta)\neq 0$ and $d_{x,\theta}\varphi(x,y,\theta)\neq 0$ on $U\times V\times\left(\mathbf{R}^N\setminus\{0\}\right)$, then we have for all $v\in\mathcal{E}'(V)$

$$\text{sing supp}\, Av\subset\Big\{x\in U : d_\theta\varphi(x,y,\theta)=0 \text{ for some } y\in\text{sing supp}\, v \text{ and some } \theta\in\mathbf{R}^N\setminus\{0\}\Big\}.$$

3.3 Pseudo-Differential Operators

Let Ω be an open subset of $\mathbf{R}^n$ and $m\in\mathbf{R}$. A *pseudo-differential operator* of order m on Ω is a Fourier integral operator of the form

$$Au(x)=\iint_{\Omega\times\mathbf{R}^n} e^{i(x-y)\cdot\xi}a(x,y,\xi)u(y)\,dy\,d\xi,\quad u\in C_0^\infty(\Omega), \tag{3.4}$$

with some $a \in S^m_{\rho,\delta}(\Omega \times \Omega \times \mathbf{R}^n)$. In other words, a pseudo-differential operator of order m is a Fourier integral operator associated with the phase function $\varphi(x, y, \xi) = (x - y) \cdot \xi$ and some amplitude $a \in S^m_{\rho,\delta}(\Omega \times \Omega \times \mathbf{R}^n)$.

We let

$$L^m_{\rho,\delta}(\Omega) = \text{the set of all pseudo-differential operators of order } m \text{ on } \Omega.$$

Applying Theorems 3.7 and 3.8 to our situation, we obtain the following:

(1) A pseudo-differential operator A maps the space $C_0^\infty(\Omega)$ continuously into the space $C^\infty(\Omega)$ and extends to a continuous linear operator $A : \mathcal{E}'(\Omega) \to \mathcal{D}'(\Omega)$.

(2) The distribution kernel K_A of a pseudo-differential operator A satisfies the condition

$$\operatorname{sing\,supp} K_A \subset \{(x, x) : x \in \Omega\},$$

that is, the kernel K_A is of class C^∞ off the diagonal $\{(x, x) : x \in \Omega\}$ in $\Omega \times \Omega$.

(3) $\operatorname{sing\,supp} Au \subset \operatorname{sing\,supp} u$, $u \in \mathcal{E}'(\Omega)$. In other words, Au is of class C^∞ whenever u is. This property is referred to as the *pseudo-local property*.

We set

$$L^{-\infty}(\Omega) = \bigcap_{m \in \mathbf{R}} L^m_{\rho,\delta}(\Omega).$$

The next theorem characterizes the class $L^{-\infty}(\Omega)$.

Theorem 3.9. *The following three conditions are equivalent:*
(i) $A \in L^{-\infty}(\Omega)$.
(ii) A is written in the form (3.4) with some $a \in S^{-\infty}(\Omega \times \Omega \times \mathbf{R}^n)$.
(iii) A is a regularizer, *or equivalently, its distribution kernel* K_A *is in the space* $C^\infty(\Omega \times \Omega)$.

We recall that a continuous linear operator $A : C_0^\infty(\Omega) \to \mathcal{D}'(\Omega)$ is said to be *properly supported* if the following two conditions are satisfied:

(a) For any compact subset K of Ω, there exists a compact subset K' of Ω such that

$$\operatorname{supp} v \subset K \implies \operatorname{supp} Av \subset K'.$$

(b) For any compact subset K' of Ω, there exists a compact subset K of Ω such that

$$\operatorname{supp} v \cap K = \emptyset \implies \operatorname{supp} Av \cap K' = \emptyset.$$

If A is properly supported, then it maps $C_0^\infty(\Omega)$ continuously into $\mathcal{E}'(\Omega)$, and further it extends to a continuous linear operator on $C^\infty(\Omega)$ into $\mathcal{D}'(\Omega)$.

The next theorem states that every pseudo-differential operator can be written as the sum of a properly supported operator and a regularizer.

Theorem 3.10. *If $A \in L^m_{\rho,\delta}(\Omega)$, then we have*

$$A = A_0 + R,$$

where $A_0 \in L^m_{\rho,\delta}(\Omega)$ is properly supported and $R \in L^{-\infty}(\Omega)$.

If $p(x,\xi) \in S^m_{\rho,\delta}(\Omega \times \mathbf{R}^n)$, then the operator $p(x,D)$, defined by

$$p(x,D)u(x) = \frac{1}{(2\pi)^n} \int_{\mathbf{R}^n} e^{ix\cdot\xi}\, p(x,\xi)\, \hat{u}(\xi)\, d\xi, \quad u \in C_0^\infty(\Omega), \tag{3.5}$$

is a pseudo-differential operator of order m on Ω, that is, $p(x,D) \in L^m_{\rho,\delta}(\Omega)$.

The next theorem asserts that every properly supported pseudo-differential operator can be reduced to the form (3.5).

Theorem 3.11. *If $A \in L^m_{\rho,\delta}(\Omega)$ is properly supported, then we have*

$$p(x,\xi) = e^{-ix\cdot\xi} A(e^{ix\cdot\xi}) \in S^m_{\rho,\delta}(\Omega \times \mathbf{R}^n),$$

and

$$A = p(x,D).$$

Furthermore, if $a(x,y,\xi) \in S^m_{\rho,\delta}(\Omega \times \Omega \times \mathbf{R}^n)$ is an amplitude for A, we have the following asymptotic expansion:

$$p(x,\xi) \sim \sum_{\alpha \geq 0} \frac{1}{\alpha!} \partial_\xi^\alpha D_y^\alpha \left(a(x,y,\xi)\right)\Big|_{y=x}.$$

The function $p(x,\xi)$ is called the complete symbol *of A.*

We extend the notion of a complete symbol to the whole space $L^m_{\rho,\delta}(\Omega)$. If $A \in L^m_{\rho,\delta}(\Omega)$, we choose a properly supported operator $A_0 \in L^m_{\rho,\delta}(\Omega)$ such that $A - A_0 \in L^{-\infty}(\Omega)$, and define

$$\sigma(A) = \text{the equivalence class of the complete symbol of } A_0 \text{ in } S^m_{\rho,\delta}(\Omega \times \mathbf{R}^n)/S^{-\infty}(\Omega \times \mathbf{R}^n).$$

In view of Theorems 3.9 and 3.10, it follows that $\sigma(A)$ does not depend on the operator A_0 chosen. The equivalence class $\sigma(A)$ is called the *complete symbol* of A. It is easy to see that the mapping

$$L^m_{\rho,\delta}(\Omega) \ni A \longmapsto \sigma(A) \in S^m_{\rho,\delta}(\Omega \times \mathbf{R}^n)/S^{-\infty}(\Omega \times \mathbf{R}^n)$$

induces an isomorphism

$$L^m_{\rho,\delta}/L^{-\infty} \longrightarrow S^m_{\rho,\delta}/S^{-\infty}.$$

We shall often identify the complete symbol $\sigma(A)$ with a representative in the class $S^m_{\rho,\delta}(\Omega \times \mathbf{R}^n)$ for notational convenience, and call any member of $\sigma(A)$ a complete symbol of A.

A pseudo-differential operator $A \in L^m_{1,0}(\Omega)$ is said to be *classical* if its complete symbol $\sigma(A)$ has a representative in the class $S^m_{cl}(\Omega \times \mathbf{R}^n)$.

We let

$L^m_{cl}(\Omega) =$ the set of all classical pseudo-differential operators of order m on Ω.

Then the mapping

$$L^m_{cl}(\Omega) \ni A \longmapsto \sigma(A) \in S^m_{cl}(\Omega \times \mathbf{R}^n)/S^{-\infty}(\Omega \times \mathbf{R}^n)$$

induces an isomorphism

$$L^m_{cl}/L^{-\infty} \longrightarrow S^m_{cl}/S^{-\infty}.$$

Also we have

$$L^{-\infty}(\Omega) = \bigcap_{m \in \mathbf{R}} L^m_{cl}(\Omega).$$

If $A \in L^m_{cl}(\Omega)$, then the principal part of $\sigma(A)$ has a canonical representative $\sigma_A(x,\xi) \in C^\infty(\Omega \times (\mathbf{R}^n \setminus \{0\}))$ which is positively homogeneous of degree m in the variable ξ. The function $\sigma_A(x,\xi)$ is called the *homogeneous principal symbol* of A.

The next two theorems assert that the class of pseudo-differential operators forms an algebra closed under the operations of composition of operators and taking the transpose or adjoint of an operator.

Theorem 3.12. *If $A \in L^m_{\rho,\delta}(\Omega)$, then its transpose A' and its adjoint A^* are both in $L^m_{\rho,\delta}(\Omega)$, and the complete symbols $\sigma(A')$ and $\sigma(A^*)$ have respectively the following asymptotic expansions:*

$$\sigma(A')(x,\xi) \sim \sum_{\alpha \geq 0} \frac{1}{\alpha!} \partial_\xi^\alpha D_x^\alpha \left(\sigma(A)(x,-\xi)\right),$$

$$\sigma(A^*)(x,\xi) \sim \sum_{\alpha \geq 0} \frac{1}{\alpha!} \partial_\xi^\alpha D_x^\alpha \left(\overline{\sigma(A)(x,\xi)}\right).$$

Theorem 3.13. *If $A \in L^{m'}_{\rho',\delta'}(\Omega)$ and $B \in L^{m''}_{\rho'',\delta''}(\Omega)$ where $0 \leq \delta' < \rho'' \leq 1$ and if one of them is properly supported, then the composition AB is in $L^{m'+m''}_{\rho,\delta}(\Omega)$ with $\rho = \min(\rho', \rho'')$, $\delta = \max(\delta', \delta'')$, and we have the following asymptotic expansion:*

$$\sigma(AB)(x,\xi) \sim \sum_{\alpha \geq 0} \frac{1}{\alpha!} \partial_\xi^\alpha \left(\sigma(A)(x,\xi)\right) \cdot D_x^\alpha \left(\sigma(B)(x,\xi)\right).$$

A pseudo-differential operator $A \in L^m_{\rho,\delta}(\Omega)$ is said to be *elliptic* of order m if its complete symbol $\sigma(A)$ is elliptic of order m. In view of Theorem 3.3, it follows that a classical pseudo-differential operator $A \in L^m_{cl}(\Omega)$ is elliptic if and only if its homogeneous principal symbol $\sigma_A(x,\xi)$ does not vanish on the space $\Omega \times (\mathbf{R}^n \setminus \{0\})$.

The next theorem states that elliptic operators are the "invertible" elements in the algebra of pseudo-differential operators.

Theorem 3.14. *An operator $A \in L^m_{\rho,\delta}(\Omega)$ is elliptic if and only if there exists a properly supported operator $B \in L^{-m}_{\rho,\delta}(\Omega)$ such that*

$$\begin{cases} AB \equiv I \mod L^{-\infty}(\Omega), \\ BA \equiv I \mod L^{-\infty}(\Omega). \end{cases}$$

Such an operator B is called a *parametrix* for A. In other words, a parametrix for A is a two-sided inverse of A modulo $L^{-\infty}(\Omega)$. We observe that a parametrix is unique modulo $L^{-\infty}(\Omega)$.

The next theorem proves the invariance of pseudo-differential operators under change of coordinates.

Theorem 3.15. *Let Ω_1, Ω_2 be two open subsets of $\mathbf{R}^n$ and $\chi : \Omega_1 \to \Omega_2$ a C^∞ diffeomorphism. If $A \in L^m_{\rho,\delta}(\Omega_1)$, where $1-\rho \leq \delta < \rho \leq 1$, then the mapping*

$$\begin{aligned} A_\chi : C_0^\infty(\Omega_2) &\longrightarrow C^\infty(\Omega_2) \\ v &\longmapsto A(v \circ \chi) \circ \chi^{-1} \end{aligned}$$

is in $L^m_{\rho,\delta}(\Omega_2)$, and we have the asymptotic expansion

$$\sigma(A_\chi)(y,\eta) \sim \sum_{\alpha \geq 0} \frac{1}{\alpha!} \left(\partial_\xi^\alpha \sigma(A)\right)(x, {}^t\chi'(x)\cdot\eta) \cdot D_z^\alpha \left(e^{ir(x,z,\eta)}\right)\Big|_{z=x} \tag{3.6}$$

with

$$r(x,z,\eta) = \langle \chi(z) - \chi(x) - \chi'(x)\cdot(z-x), \eta \rangle .$$

Here $x = \chi^{-1}(y)$, $\chi'(x)$ is the derivative of χ at x and ${}^t\chi'(x)$ its transpose.

Remark 3.16. Formula (3.6) shows that

$$\sigma(A_\chi)(y,\eta) \equiv \sigma(A)\left(x, {}^t\chi'(x)\cdot\eta\right) \mod S^{m-(\rho-\delta)}_{\rho,\delta}.$$

Note that the mapping

$$\Omega_2 \times \mathbf{R}^n \ni (y,\eta) \longmapsto \left(x, {}^t\chi'(x)\cdot\eta\right) \in \Omega_1 \times \mathbf{R}^n$$

is just a transition map of the cotangent bundle $T^*(\mathbf{R}^n)$. This implies that the principal symbol $\sigma_m(A)$ of $A \in L^m_{\rho,\delta}(\mathbf{R}^n)$ can be invariantly defined on $T^*(\mathbf{R}^n)$ when $1-\rho \leq \delta < \rho \leq 1$.

A differential operator of order m with C^∞ coefficients on Ω is continuous on $H^{s,p}_{loc}(\Omega)$ (resp. $B^{s,p}_{loc}(\Omega)$) into $H^{s-m,p}_{loc}(\Omega)$ (resp. $B^{s-m,p}_{loc}(\Omega)$) for all $s \in \mathbf{R}$. This result extends to pseudo-differential operators (cf. [Bo, Theorem 1]):

Theorem 3.17. *Every properly supported operator $A \in L^m_{1,\delta}(\Omega)$, $0 \leq \delta < 1$, extends to continuous linear operators*

$$A : H^{s,p}_{loc}(\Omega) \longrightarrow H^{s-m,p}_{loc}(\Omega),$$
$$A : B^{s,p}_{loc}(\Omega) \longrightarrow B^{s-m,p}_{loc}(\Omega)$$

for all $s \in \mathbf{R}$.

Now we define the concept of a pseudo-differential operator on a manifold, and transfer all the machinery of pseudo-differential operators to manifolds. Let M be an n-dimensional *compact* C^∞ manifold without boundary. Theorem 3.15 leads us to the following:

Definition 3.18. Let $1 - \rho \leq \delta < \rho \leq 1$. A continuous linear operator $A : C^\infty(M) \to C^\infty(M)$ is called a *pseudo-differential operator* of order $m \in \mathbf{R}$ if it satisfies the following two conditions:

(i) The distribution kernel of A is of class C^∞ off the diagonal $\{(x, x) : x \in M\}$ in $M \times M$.

(ii) For any chart (U, χ) on M, the mapping

$$A_\chi : C_0^\infty(\chi(U)) \longrightarrow C^\infty(\chi(U))$$
$$u \longmapsto A(u \circ \chi) \circ \chi^{-1}$$

belongs to the class $L^m_{\rho,\delta}(\chi(U))$.

We let

$$L^m_{\rho,\delta}(M) = \text{the set of all pseudo-differential operators of order } m \text{ on } M,$$

and set

$$L^{-\infty}(M) = \bigcap_{m \in \mathbf{R}} L^m_{\rho,\delta}(M).$$

Some results about pseudo-differential operators on $\mathbf{R}^n$ stated above are also true for pseudo-differential operators on M. In fact, pseudo-differential operators on M are defined to be locally pseudo-differential operators on $\mathbf{R}^n$.

For example, we have the following results:

(1) A pseudo-differential operator $A : C^\infty(M) \to C^\infty(M)$ extends to a continuous linear operator $A : \mathcal{D}'(M) \to \mathcal{D}'(M)$.

(2) $\text{sing supp}\, Au \subset \text{sing supp}\, u$, $u \in \mathcal{D}'(M)$.

(3) A continuous linear operator $A : C^\infty(M) \to \mathcal{D}'(M)$ is a regularizer if and only if it is in $L^{-\infty}(M)$.

(4) The class $L^m_{\rho,\delta}(M)$ is stable under the operations of composition of operators and taking the transpose or adjoint of an operator.

(5) A pseudo-differential operator $A \in L^m_{1,\delta}(M)$, $0 \leq \delta < 1$, extends to continuous linear operators $A : H^{s,p}(M) \to H^{s-m,p}(M)$ and $A : B^{s,p}(M) \to B^{s-m,p}(M)$ for all $s \in \mathbf{R}$.

A pseudo-differential operator $A \in L^m_{1,0}(M)$ is said to be *classical* if, for any chart (U, χ) on M, the mapping $A_\chi : C_0^\infty(\chi(U)) \to C^\infty(\chi(U))$ belongs to the class $L^m_{cl}(\chi(U))$.

We let

$L^m_{cl}(M)$ = the set of all classical pseudo-differential operators of order m on M.

We observe that

$$L^{-\infty}(M) = \bigcap_{m \in \mathbf{R}} L^m_{cl}(M).$$

Let $A \in L^m_{cl}(M)$. If (U, χ) is a chart on M, there is associated a homogeneous principal symbol $\sigma_{A_\chi} \in C^\infty(\chi(U) \times (\mathbf{R}^n \setminus \{0\}))$. In view of Remark 3.16, by smoothly patching together the functions σ_{A_χ}, one can obtain a C^∞ function $\sigma_A(x, \xi)$ on $T^*(M) \setminus \{0\} = \{(x, \xi) \in T^*(M) : \xi \neq 0\}$, which is positively homogeneous of degree m in the variable ξ. The function σ_A is called the *homogeneous principal symbol* of A.

A classical pseudo-differential operator $A \in L^m_{cl}(M)$ is said to be *elliptic* of order m if its homogeneous principal symbol $\sigma_A(x, \xi)$ does not vanish on the bundle $T^*(M) \setminus \{0\}$ of non-zero cotangent vectors.

Then we have the following:

(6) An operator $A \in L^m_{cl}(M)$ is elliptic if and only if there exists a parametrix $B \in L^{-m}_{cl}(M)$ for A:

$$\begin{cases} AB \equiv I \mod L^{-\infty}(M), \\ BA \equiv I \mod L^{-\infty}(M). \end{cases}$$

Let Ω be an open subset of $\mathbf{R}^n$. A properly supported pseudo-differential operator A on Ω is said to be *hypoelliptic* if it satisfies the condition

$$\text{sing supp}\, u = \text{sing supp}\, Au\,, \quad u \in \mathcal{D}'(\Omega).$$

For example, Theorem 3.14 tells us that elliptic operators are hypoelliptic. We remark that this notion may be transferred to manifolds.

The following criterion for hypoellipticity is due to Hörmander (cf. [Ho, Chapter XXII, Theorem 22.1.3]):

Theorem 3.19. *Let $A = p(x, D) \in L^m_{\rho,\delta}(\Omega)$ be properly supported. Assume that, for any compact $K \subset \Omega$ and any multi-indices α, β, there exist constants $C_{K,\alpha,\beta} > 0$, $C_K > 0$ and $\mu \in \mathbf{R}$ such that we have for all $x \in K$ and $|\xi| \geq C_K$*

$$\left|D_\xi^\alpha D_x^\beta p(x, \xi)\right| \leq C_{K,\alpha,\beta} |p(x, \xi)| (1 + |\xi|)^{-\rho|\alpha| + \delta|\beta|}, \tag{3.7a}$$

$$|p(x, \xi)|^{-1} \leq C_K (1 + |\xi|)^\mu. \tag{3.7b}$$

Then there exists a parametrix $B \in L^\mu_{\rho,\delta}(\Omega)$ for A.

CHAPTER IV

L^p APPROACH TO ELLIPTIC BOUNDARY VALUE PROBLEMS

In this chapter we study elliptic boundary value problems in the framework of Sobolev spaces of L^p-style, by using the L^p theory of pseudo-differential operators. For more thorough treatments of this subject, the reader might refer to Seeley [Se3], Taylor [Ty], Chazarain and Piriou [CP] and also Taira [Ta2] (L^2 version).

4.1 The Dirichlet Problem

In this section, we shall consider the Dirichlet problem in the framework of Sobolev spaces of L^p style which is a generalization of the classical potential approach to the Dirichlet problem.

Let Ω be a bounded domain of Euclidean space $\mathbf{R}^n$ with C^∞ boundary Γ. Its closure $\overline{\Omega} = \Omega \cup \Gamma$ is an n-dimensional compact C^∞ manifold with boundary. One may assume that $\overline{\Omega}$ is the closure of a relatively compact open subset Ω of an n-dimensional compact C^∞ manifold M without boundary in which Ω has a C^∞ boundary Γ (see Figure 3.1). Let μ be a strictly positive density on M and ω a strictly positive density on Γ.

We let

$$A = \sum_{i=1}^{n} \frac{\partial}{\partial x_i} \left(\sum_{j=1}^{n} a^{ij}(x) \frac{\partial}{\partial x_j} \right) + \sum_{i=1}^{n} b^i(x) \frac{\partial}{\partial x_i} + c(x)$$

be a second-order *elliptic* differential operator with real coefficients such that:

(1) $a^{ij} \in C^\infty(M)$, $a^{ij} = a^{ji}$ and there exists a constant $a_0 > 0$ such that

$$\sum_{i,j=1}^{n} a^{ij}(x)\xi_i\xi_j \geq a_0|\xi|^2 \quad \text{on } T^*(M).$$

Here $T^*(M)$ is the cotangent bundle of M.

(2) $b^i \in C^\infty(M)$.

(3) $c \in C^\infty(M)$ and $c \leq 0$ in M.

Further, for simplicity, assume that

(4.1) The function c does not vanish *identically* on M.

Typeset by $\mathcal{A}\mathcal{M}\mathcal{S}$-TEX

First we consider the Dirichlet problem:

$$\text{(D)} \qquad \begin{cases} Au = f & \text{in } \Omega, \\ u|_\Gamma = \varphi & \text{on } \Gamma. \end{cases}$$

The next theorem states the existence of a volume potential for A, which plays the same role for A as the Newtonian potential plays for the Laplacian (cf. [Se2, Theorem 1], [Ta1, Theorem 8.2.1]):

Theorem 4.1. *(i) The operator $A : C^\infty(M) \longrightarrow C^\infty(M)$ is bijective, and its inverse Q is a classical elliptic pseudo-differential operator of order -2 on M.*

(ii) The operators A and Q extend respectively to isomorphisms

$$A : H^{s,p}(M) \longrightarrow H^{s-2,p}(M),$$
$$Q : H^{s-2,p}(M) \longrightarrow H^{s,p}(M)$$

for all $s \in \mathbf{R}$ and $1 < p < \infty$, which are still inverses of each other.

Next we construct a surface potential for A, which is a generalization of the classical Poisson kernel for the Laplacian.

We let

$$Kv = Q(v \otimes \delta)|_\Gamma, \quad v \in C^\infty(\Gamma),$$

where $v \otimes \delta$ is a distribution on M defined by

$$\langle v \otimes \delta, \varphi \cdot \mu \rangle = \langle v, \varphi|_\Gamma \cdot \omega \rangle, \quad \varphi \in C^\infty(M).$$

Then we have the following (cf. [Ta1, Theorem 8.2.2]):

Theorem 4.2. *(i) The operator K is a classical elliptic pseudo-differential operator of order -1 on Γ.*

(ii) The operator $K : C^\infty(\Gamma) \longrightarrow C^\infty(\Gamma)$ is bijective, and its inverse L is a classical elliptic pseudo-differential operator of first order on Γ. Furthermore, the operators K and L extend respectively to isomorphisms

$$K : B^{\sigma,p}(\Gamma) \longrightarrow B^{\sigma+1,p}(\Gamma),$$
$$L : B^{\sigma+1,p}(\Gamma) \longrightarrow B^{\sigma,p}(\Gamma)$$

for all $\sigma \in \mathbf{R}$ and $1 < p < \infty$, which are still inverses of each other.

Now we let

$$P\varphi = Q(L\varphi \otimes \delta)|_\Omega, \quad \varphi \in C^\infty(\Gamma).$$

Then the operator P maps $C^\infty(\Gamma)$ continuously into $C^\infty(\overline{\Omega})$, and extends to a continuous linear operator

$$P : B^{s-1/p,p}(\Gamma) \longrightarrow H^{s,p}(\Omega)$$

for all $s \in \mathbf{R}$ and $1 < p < \infty$. Further we have for all $\varphi \in B^{s-1/p,p}(\Gamma)$

$$\begin{cases} AP\varphi = AQ(L\varphi \otimes \delta)|_\Omega = (L\varphi \otimes \delta)|_\Omega = 0 & \text{in } \Omega, \\ P\varphi|_\Gamma = KL\varphi = \varphi & \text{on } \Gamma. \end{cases}$$

The operator P is called the *Poisson operator.*

We let

$$N(A, s, p) = \{u \in H^{s,p}(\Omega) : \ Au = 0 \text{ in } \Omega\}.$$

Since the injection $H^{s,p}(\Omega) \longrightarrow \mathcal{D}'(\Omega)$ is continuous, it follows that the space $N(A, s, p)$ is a closed subspace of $H^{s,p}(\Omega)$; hence it is a Banach space.

Then we have the following (cf. [Se3, Theorems 5 and 6]):

Theorem 4.3. *The Poisson operator P maps the space $B^{s-1/p,p}(\Gamma)$ isomorphically onto the space $N(A, s, p)$ for all $s \in \mathbf{R}$ and $1 < p < \infty$.*

We remark that the spaces $N(A, s, p)$ and $B^{s-1/p,p}(\Gamma)$ are isomorphic in such a way that

$$N(A, s, p) \xrightarrow{\gamma_0} B^{s-1/p,p}(\Gamma),$$
$$N(A, s, p) \underset{P}{\longleftarrow} B^{s-1/p,p}(\Gamma).$$

Combining Theorems 4.1 and 4.3, we can obtain the following (cf. [ADN], [LM]):

Theorem 4.4. *Let $s > 1/p$ where $1 < p < \infty$. The Dirichlet problem (D) has a unique solution u in the space $H^{s,p}(\Omega)$ for any $f \in H^{s-2,p}(\Omega)$ and any $\varphi \in B^{s-1/p,p}(\Gamma)$.*

Next we consider the Neumann problem:

$$\text{(N)} \qquad \begin{cases} Au = f & \text{in } \Omega, \\ \frac{\partial u}{\partial \nu}\Big|_\Gamma = \varphi & \text{on } \Gamma. \end{cases}$$

Then we have the following (cf. [ADN], [LM]):

Theorem 4.5. *Let $s > 1 + 1/p$ where $1 < p < \infty$. The Neumann problem (N) has a unique solution u in the space $H^{s,p}(\Omega)$ for any $f \in H^{s-2,p}(\Omega)$ and any $\varphi \in B^{s-1-1/p,p}(\Gamma)$.*

4.2 Formulation of a Boundary Value Problem

Let $s > 1 + 1/p$ where $1 < p < \infty$. If $u \in H^{s,p}(\Omega)$, we can define its traces $\gamma_0 u$ and $\gamma_1 u$ respectively by the formulas

$$\begin{cases} \gamma_0 u = u|_\Gamma, \\ \gamma_1 u = \frac{\partial u}{\partial \nu}\Big|_\Gamma, \end{cases}$$

and let

$$\gamma u = \{\gamma_0 u, \gamma_1 u\}.$$

Then we have the following (cf. [St]):

Theorem 4.6 (The trace theorem). *The trace map*

$$\gamma : H^{s,p}(\Omega) \longrightarrow B^{s-1/p,p}(\Gamma) \times B^{s-1-1/p,p}(\Gamma)$$

is continuous and surjective, for $s > 1 + 1/p$ where $1 < p < \infty$.

We introduce a subspace of $B^{s-1-1/p,p}(\Gamma)$ which is a suitable tool to investigate the boundary condition

$$Bu = a\frac{\partial u}{\partial \boldsymbol{\nu}} + bu\Big|_{\Gamma} = a\gamma_1 u + b\gamma_0 u, \quad u \in H^{s,p}(\Omega). \tag{4.2}$$

We let

$$B_*^{s-1-1/p,p}(\Gamma) = \left\{\varphi = a\varphi_1 + b\varphi_2 : \varphi_1 \in B^{s-1-1/p,p}(\Gamma),\ \varphi_2 \in B^{s-1/p,p}(\Gamma)\right\},$$

and define a norm

$$|\varphi|^*_{s-1-1/p,p} = \inf\left\{|\varphi_1|_{s-1-1/p,p} + |\varphi_2|_{s-1/p,p} :\ \varphi = a\varphi_1 + b\varphi_2\right\}.$$

Then we have the following:

Lemma 4.7. *The space $B_*^{s-1-1/p,p}(\Gamma)$ is a Banach space with respect to the norm $|\cdot|^*_{s-1-1/p,p}$.*

Proof. (i) First we verify that the quantity $|\cdot|^*_{s-1-1/p,p}$ satisfies the axioms of a norm.

(a) If $|\varphi|^*_{s-1-1/p,p} = 0$, then one can find a sequence $\{\varphi_1^{(\ell)}\}$ in the space $B^{s-1-1/p,p}(\Gamma)$ and a sequence $\{\varphi_2^{(\ell)}\}$ in the space $B^{s-1/p,p}(\Gamma)$ such that

$$\varphi = a\varphi_1^{(\ell)} + b\varphi_2^{(\ell)},$$

with

$$|\varphi_1^{(\ell)}|_{s-1-1/p,p} < \frac{1}{\ell},$$
$$|\varphi_2^{(\ell)}|_{s-1/p,p} < \frac{1}{\ell}.$$

Hence we have with a constant $C > 0$

$$\begin{aligned} |\varphi|_{s-1-1/p,p} &\le |a\varphi_1^{(\ell)}|_{s-1-1/p,p} + |b\varphi_2^{(\ell)}|_{s-1-1/p,p} \\ &\le C\left(|\varphi_1^{(\ell)}|_{s-1-1/p,p} + |\varphi_2^{(\ell)}|_{s-1/p,p}\right) \\ &\le \frac{2C}{\ell} \longrightarrow 0 \text{ as } \ell \to \infty, \end{aligned}$$

so that

$$\varphi = 0.$$

(b) We have for all $\lambda \in \mathbf{C}$

$$\begin{aligned} |\lambda\varphi|^*_{s-1-1/p,p} &= \inf\left\{|\lambda\varphi_1|_{s-1-1/p,p} + |\lambda\varphi_2|_{s-1/p,p} :\ \lambda\varphi = a\lambda\varphi_1 + b\lambda\varphi_2\right\} \\ &= |\lambda| \inf\left\{|\varphi_1|_{s-1-1/p,p} + |\varphi_2|_{s-1/p,p} :\ \varphi = a\varphi_1 + b\varphi_2\right\} \\ &= |\lambda|\, |\varphi|^*_{s-1-1/p,p}. \end{aligned}$$

(c) If we have

$$\begin{aligned} \varphi &= a\varphi_1 + b\varphi_2 \text{ with } \varphi_1 \in B^{s-1-1/p,p}(\Gamma),\ \varphi_2 \in B^{s-1/p,p}(\Gamma), \\ \psi &= a\psi_1 + b\psi_2 \text{ with } \psi_1 \in B^{s-1-1/p,p}(\Gamma),\ \psi_2 \in B^{s-1/p,p}(\Gamma), \end{aligned}$$

then it follows that

$$\varphi + \psi = a(\varphi_1 + \psi_1) + b(\varphi_2 + \psi_2),$$

so that

$$\begin{aligned} &|\varphi + \psi|^*_{s-1-1/p,p} \\ &\le |\varphi_1 + \psi_1|_{s-1-1/p,p} + |\varphi_2 + \psi_2|_{s-1/p,p} \\ &\le \left(|\varphi_1|_{s-1-1/p,p} + |\varphi_2|_{s-1/p,p}\right) + \left(|\psi_1|_{s-1-1/p,p} + |\psi_2|_{s-1/p,p}\right). \end{aligned}$$

This proves that

$$|\varphi + \psi|^*_{s-1-1/p,p} \le |\varphi|^*_{s-1-1/p,p} + |\psi|^*_{s-1-1/p,p}.$$

(ii) To prove the completeness of the space $B_*^{s-1-1/p,p}(\Gamma)$, it suffices to show that the condition

$$\sum_{\ell=1}^{\infty} |\varphi^{(\ell)}|^*_{s-1-1/p,p} < \infty \tag{4.3}$$

implies that there is an element $\varphi \in B_*^{s-1-1/p,p}(\Gamma)$ such that

$$\left| \sum_{\ell=1}^{N} \varphi^{(\ell)} - \varphi \right|^*_{s-1-1/p,p} \longrightarrow 0 \text{ as } N \to \infty.$$

By condition (4.3), one can find a sequence $\{\varphi_1^{(\ell)}\}$ in $B^{s-1-1/p,p}(\Gamma)$ and a sequence $\{\varphi_2^{(\ell)}\}$ in $B^{s-1/p,p}(\Gamma)$ such that

$$\varphi = a\varphi_1^{(\ell)} + b\varphi_2^{(\ell)},$$

with

$$|\varphi_1^{(\ell)}|_{s-1-1/p,p} + |\varphi_2^{(\ell)}|_{s-1/p,p} \leq |\varphi^{(\ell)}|^*_{s-1-1/p,p} + \frac{1}{2^\ell}.$$

Then we have

$$\sum_{\ell=1}^{\infty} |\varphi_1^{(\ell)}|_{s-1-1/p,p}, \ \sum_{\ell=1}^{\infty} |\varphi_2^{(\ell)}|_{s-1/p,p} \leq \sum_{\ell=1}^{\infty} \left(|\varphi^{(\ell)}|^*_{s-1-1/p,p} + \frac{1}{2^\ell} \right)$$
$$\leq \sum_{\ell=1}^{\infty} |\varphi^{(\ell)}|^*_{s-1-1/p,p} + 1.$$

Hence, since the spaces $B^{s-1-1/p,p}(\Gamma)$ and $B^{s-1/p,p}(\Gamma)$ are complete, one can find two elements $\varphi_1 \in B^{s-1-1/p,p}(\Gamma)$ and $\varphi_2 \in B^{s-1/p,p}(\Gamma)$ such that

$$\left| \sum_{\ell=1}^{N} \varphi_1^{(\ell)} - \varphi_1 \right|_{s-1-1/p,p} \longrightarrow 0 \text{ as } N \to \infty,$$
$$\left| \sum_{\ell=1}^{N} \varphi_2^{(\ell)} - \varphi_2 \right|_{s-1/p,p} \longrightarrow 0 \text{ as } N \to \infty.$$

Therefore, letting

$$\varphi = a\varphi_1 + b\varphi_2 \in B_*^{s-1-1/p,p}(\Gamma),$$

we obtain that, as $N \to \infty$,

$$\left| \sum_{\ell=1}^{N} \varphi^{(\ell)} - \varphi \right|^*_{s-1-1/p,p}$$
$$\leq \left| \sum_{\ell=1}^{N} \varphi_1^{(\ell)} - \varphi_1 \right|_{s-1-1/p,p} + \left| \sum_{\ell=1}^{N} \varphi_2^{(\ell)} - \varphi_2 \right|_{s-1/p,p} \longrightarrow 0.$$

This completes the proof of Lemma 4.7. □

Furthermore it is easy to verify the following:

Proposition 4.8. *The mapping*

$$B : H^{s,p}(\Omega) \longrightarrow B_*^{s-1-1/p,p}(\Gamma)$$

is continuous and surjective, for $s > 1 + 1/p$ *where* $1 < p < \infty$.

Now we can formulate our boundary value problem for (A, B) as follows: Given functions $f \in H^{s-2,p}(\Omega)$ and $\varphi \in B_*^{s-1-1/p,p}(\Gamma)$, find a function $u \in H^{s,p}(\Omega)$ such that

$$(*) \qquad \begin{cases} Au = f & \text{in } \Omega, \\ Bu = \varphi & \text{on } \Gamma. \end{cases}$$

4.3 Reduction to the Boundary

In this section, we shall show that problem $(*)$ can be reduced to the study of a pseudo-differential operator on the boundary.

Let f be an arbitrary element of $H^{s-2,p}(\Omega)$, and φ an arbitrary element of $B_*^{s-1-1/p,p}(\Gamma)$ such that

$$\varphi = a\varphi_1 + b\varphi_2, \quad \varphi_1 \in B^{s-1-1/p,p}(\Gamma), \ \varphi_2 \in B^{s-1/p,p}(\Gamma),$$

where $1 < p < \infty$ and $s > 1 + 1/p$. First we consider the following Neumann problem:

$$\text{(N)} \qquad \begin{cases} Av = f & \text{in } \Omega, \\ \frac{\partial v}{\partial \nu}\Big|_\Gamma = \varphi_1 & \text{on } \Gamma. \end{cases}$$

By Theorem 4.5, one can find a unique solution $v \in H^{s,p}(\Omega)$ of problem (N). Then it is easy to see that a function u in $H^{s,p}(\Omega)$ is a solution of problem

$$(*) \qquad \begin{cases} Au = f & \text{in } \Omega, \\ Bu = \varphi & \text{on } \Gamma \end{cases}$$

if and only if $w = u - v \in H^{s,p}(\Omega)$ is a solution of the problem

$$(*') \qquad \begin{cases} Aw = 0 & \text{in } \Omega, \\ Bw = \varphi - Bv = b(\varphi_2 - v|_\Gamma) & \text{on } \Gamma. \end{cases}$$

But Theorem 4.3 tells us that the spaces $N(A, s, p)$ and $B^{s-1/p,p}(\Gamma)$ are isomorphic in such a way that

$$N(A, s, p) \xrightarrow{\gamma_0} B^{s-1/p,p}(\Gamma),$$
$$N(A, s, p) \xleftarrow[P]{} B^{s-1/p,p}(\Gamma).$$

Therefore, we find that $w \in H^{s,p}(\Omega)$ is a solution of problem $(*')$ if and only if $\psi \in B^{s-1/p,p}(\Gamma)$ is a solution of the equation

$$(\dagger) \qquad BP\psi = b(\varphi_2 - \gamma_0 v) \quad \text{on } \Gamma.$$

Here $\psi = \gamma_0 w$, or equivalently, $w = P\psi$.

Summing up, we obtain the following:

Proposition 4.9. *Let $1 < p < \infty$ and $s > 1 + 1/p$. For functions $f \in H^{s-2,p}(\Omega)$ and $\varphi \in B_*^{s-1-1/p,p}(\Gamma)$, there exists a solution $u \in H^{s,p}(\Omega)$ of problem $(*)$ if and only if there exists a solution $\psi \in B^{s-1/p,p}(\Gamma)$ of equation $(\dagger)$.*

We remark that equation $(\dagger)$ is a generalization of the classical *Fredholm integral equation.*

We let

$$\begin{aligned} T : C^\infty(\Gamma) &\longrightarrow C^\infty(\Gamma) \\ \varphi &\longmapsto BP\varphi. \end{aligned}$$

Then we have by condition (4.2)

$$T = a\Pi + b,$$

where

$$\Pi\varphi = \gamma_1 P\varphi = \frac{\partial}{\partial \boldsymbol{\nu}}(P\varphi)\Big|_\Gamma .$$

It is known (cf. [Ho, Chapter XX], [Se3], [RS, Chapter 3]) that the operator Π is a classical pseudo-differential operator of first order on Γ; hence the operator T is a classical pseudo-differential operator of first order on the boundary Γ. We remark that the operator $T : C^\infty(\Gamma) \longrightarrow C^\infty(\Gamma)$ extends to a continuous linear operator $T : B^{\sigma,p}(\Gamma) \longrightarrow B^{\sigma-1,p}(\Gamma)$ for all $\sigma \in \mathbf{R}$.

Consequently, Proposition 4.9 asserts that problem $(*)$ can be reduced to the study of the pseudo-differential operator T on the boundary Γ. We shall formulate this fact more precisely in terms of functional analysis.

We associate with problem $(*)$ a continuous linear operator

$$\mathcal{A} : H^{s,p}(\Omega) \longrightarrow H^{s-2,p}(\Omega) \times B_*^{s-1-1/p,p}(\Gamma)$$

as follows.

(a) The domain $D(\mathcal{A})$ of $\mathcal{A}$ is the space $H^{s,p}(\Omega)$.

(b) $\mathcal{A}u = \{Au, Bu\}$, $u \in D(\mathcal{A})$.

Similarly, we associate with equation (†) a linear operator

$$\mathcal{T} : B^{s-1/p,p}(\Gamma) \longrightarrow B^{s-1/p,p}(\Gamma)$$

as follows.

(α) The domain $D(\mathcal{T})$ of $\mathcal{T}$ is the space

$$D(\mathcal{T}) = \left\{\varphi \in B^{s-1/p,p}(\Gamma) : T\varphi \in B^{s-1/p,p}(\Gamma)\right\}.$$

(β) $\mathcal{T}\varphi = T\varphi$, $\varphi \in D(\mathcal{T})$.

We remark that the operator $\mathcal{T}$ is a densely defined, closed operator, since the operator $T : B^{s-1/p,p}(\Gamma) \longrightarrow B^{s-1-1/p,p}(\Gamma)$ is continuous and since the domain $D(\mathcal{T})$ contains the space $C^\infty(\Gamma)$.

Then Proposition 4.9 can be reformulated in the following form (cf. [Ta2, Section 8.3]):

Theorem 4.10. *(i) The null space $N(\mathcal{A})$ of $\mathcal{A}$ has finite dimension if and only if the null space $N(\mathcal{T})$ of $\mathcal{T}$ has finite dimension, and we have*

$$\dim N(\mathcal{A}) = \dim N(\mathcal{T}).$$

(ii) The range $R(\mathcal{A})$ of $\mathcal{A}$ is closed if and only if the range $R(\mathcal{T})$ of $\mathcal{T}$ is closed; and $R(\mathcal{A})$ has finite codimension if and only if $R(\mathcal{T})$ has finite codimension, and we have

$$\operatorname{codim} R(\mathcal{A}) = \operatorname{codim} R(\mathcal{T}).$$

(iii) The operator $\mathcal{A}$ is a Fredholm operator if and only if the operator $\mathcal{T}$ is a Fredholm operator, and we have

$$\operatorname{ind} \mathcal{A} = \operatorname{ind} \mathcal{T}.$$

Here we recall that a densely defined, closed linear operator T from a Banach space X into a Banach space Y is called a *Fredholm operator* if it satisfies the following three conditions:

(a) The null space $N(T)$ of T has finite dimension.
(b) The range $R(T)$ of T is closed in Y.
(c) The range $R(T)$ has finite codimension in Y.

In this case, the *index* of T is defined by the formula

$$\operatorname{ind} T = \dim N(T) - \operatorname{codim} R(T).$$

Furthermore the next theorem states that the operator $\mathcal{A}$ has the regularity property if and only if the operator $\mathcal{T}$ has.

Theorem 4.11. *Let $1 < p < \infty$ and $s > 1 + 1/p$. The following two conditions are equivalent:*

(4.4)
$$u \in L^p(\Omega),\ Au \in H^{s-2,p}(\Omega),\ Bu \in B_*^{s-1-1/p,p}(\Gamma) \implies u \in H^{s,p}(\Omega);$$

$$\varphi \in B^{-1/p,p}(\Gamma),\ T\varphi \in B^{s-1/p,p}(\Gamma) \implies \varphi \in B^{s-1/p,p}(\Gamma). \tag{4.5}$$

Proof. (i) (4.4) $\Rightarrow$ (4.5): Assume that

$$\varphi \in B^{-1/p,p}(\Gamma) \quad \text{and} \quad T\varphi \in B^{s-1/p,p}(\Gamma).$$

Then, letting $u = P\varphi$, we obtain that

$$u \in L^p(\Omega),\ Au = 0 \quad \text{and} \quad Bu = T\varphi \in B^{s-1/p,p}(\Gamma).$$

Hence it follows from condition (4.4) that

$$u \in H^{s,p}(\Omega),$$

so that by Theorem 4.6

$$\varphi = \gamma_0 u \in B^{s-1/p,p}(\Gamma).$$

(ii) (4.5) $\Rightarrow$ (4.4): Conversely, we assume that

$$u \in L^p(\Omega), \; Au \in H^{s-2,p}(\Omega), \; Bu \in B_*^{s-1-1/p,p}(\Gamma)$$

where

$$Bu = a\varphi_1 + b\varphi_2,$$

with

$$\varphi_1 \in B^{s-1-1/p,p}(\Gamma), \quad \varphi_2 \in B^{s-1/p,p}(\Gamma).$$

Then the function u can be decomposed as follows:

$$u = v + w.$$

Here $v \in H^{s,p}(\Omega)$ is a solution of the Neumann problem

$$\begin{cases} Av = Au & \text{in } \Omega, \\ \frac{\partial v}{\partial \nu}\big|_\Gamma = \varphi_1 & \text{on } \Gamma, \end{cases}$$

and so

$$w = u - v \in N(A, 0, p).$$

But Theorem 4.3 tells us that the function w can be written as

$$w = P\varphi, \quad \varphi = \gamma_0 w \in B^{-1/p,p}(\Gamma).$$

Hence we have by Theorem 4.6

$$T\varphi = BP\varphi = Bu - Bv = b(\varphi_2 - v|_\Gamma) \in B^{s-1/p,p}(\Gamma).$$

Thus it follows from condition (4.5) that

$$\varphi \in B^{s-1/p,p}(\Gamma),$$

so that again by Theorem 4.3

$$w = P\varphi \in H^{s,p}(\Omega).$$

This proves that

$$u = v + w \in H^{s,p}(\Omega).$$

The proof of Theorem 4.11 is complete. $\square$

The next theorem states that *a priori* estimates for $\mathcal{A}$ are entirely equivalent to the corresponding *a priori* estimates for T:

Theorem 4.12. *The following two conditions are equivalent:*

(4.6)
$$\|u\|_{s,p} \le C\left(\|Au\|_{s-2,p} + |Bu|^*_{s-1-1/p,p} + \|u\|_p\right), \quad u \in D(\mathcal{A});$$

(4.7)
$$|\varphi|_{s-1/p,p} \le C\left(|\mathcal{T}\varphi|_{s-1/p,p} + |\varphi|_{-1/p,p}\right), \quad \varphi \in D(\mathcal{T}).$$

Here and in the following the letter C denotes a generic positive constant.

Proof. (i) (4.7) $\Rightarrow$ (4.6): We decompose a function $u \in D(\mathcal{A})$ as in the proof of Theorem 4.11:

$$u = v + w,$$

where $Bu = a\varphi_1 + b\varphi_2$ and

$$\begin{cases} Av = Au & \text{in } \Omega, \\ \frac{\partial v}{\partial \nu}\big|_\Gamma = \varphi_1 & \text{on } \Gamma. \end{cases}$$

Then we have by Theorem 4.5

$$\|v\|_{s,p} \le C\left(\|Au\|_{s-2,p} + |\varphi_1|_{s-1-1/p,p}\right). \tag{4.8}$$

Further, applying estimate (4.7) to the function $\gamma_0 w$, we obtain that

$$\begin{aligned} |\gamma_0 w|_{s-1/p,p} &\le C\left(|\mathcal{T}(\gamma_0 w)|_{s-1/p,p} + |\gamma_0 w|_{-1/p,p}\right) \\ &= C\left(|Bw|_{s-1/p,p} + |\gamma_0 w|_{-1/p,p}\right) \\ &\le C\left(|\varphi_2|_{s-1/p,p} + |\gamma_0 v|_{s-1/p,p} + |\gamma_0 w|_{-1/p,p}\right) \\ &\le C\left(|\varphi_2|_{s-1/p,p} + \|v\|_{s,p} + |\gamma_0 w|_{-1/p,p}\right). \end{aligned}$$

In view of Theorem 4.3, this gives that

$$\begin{aligned} \|w\|_{s,p} &\le C\left(|\varphi_2|_{s-1/p,p} + \|v\|_{s,p} + \|w\|_p\right) \\ &\le C\left(|\varphi_2|_{s-1/p,p} + \|u\|_p + \|v\|_{s,p}\right). \end{aligned} \tag{4.9}$$

Thus, combining estimates (4.8) and (4.9), we obtain that

$$\|w\|_{s,p} \le C\left(\|Au\|_{s-2,p} + |\varphi_1|_{s-1-1/p,p} + |\varphi_2|_{s-1/p,p} + \|u\|_p\right). \tag{4.10}$$

Therefore, the desired estimate (4.6) follows from estimates (4.8) and (4.10), since $u = v + w$.

(ii) (4.6) $\Rightarrow$ (4.7): Conversely, taking $u = P\varphi$ with $\varphi \in D(\mathcal{T})$ in estimate (4.6), we obtain that

$$\begin{aligned} \|P\varphi\|_{s,p} &\le C\left(|\mathcal{T}\varphi|^*_{s-1-1/p,p} + \|P\varphi\|_p\right) \\ &\le C\left(|\mathcal{T}\varphi|_{s-1/p,p} + \|P\varphi\|_p\right). \end{aligned} \tag{4.11}$$

But Theorem 4.3 tells us that the operator P maps the space $B^{\sigma-1/p,p}(\Gamma)$ isomorphically onto the space $N(A, \sigma, p)$ for all $\sigma \in \mathbf{R}$. Thus estimate (4.7) follows from estimate (4.11). $\square$

4.4 Operator Π

The next theorem summarizes two important properties of Π as a pseudo-differential operator:

Theorem 4.13. *(i) The operator Π is* strongly elliptic, *that is, there exist constants $c_1 > 0$ and $c_2 > 0$ such that for all $\varphi \in C^\infty(\Gamma)$*

$$\operatorname{Re} \int_\Gamma \Pi\varphi \cdot \overline{\varphi}\, d\sigma \geq c_1 |\varphi|^2_{1/2,2} - c_2 |\varphi|^2_{-1/2,2}. \tag{4.12}$$

(ii) Let $p_1(x', \xi') + \sqrt{-1}\, q_1(x', \xi')$ be the principal symbol of the operator Π. Then there exists a constant $c_0 > 0$ such that

$$p_1(x', \xi') \geq c_0 |\xi'| \quad \text{on } T^*(\Gamma) \setminus \{0\}. \tag{4.13}$$

Here $T^*(\Gamma) \setminus \{0\}$ is the bundle of non-zero cotangent vectors, and $|\xi'|$ is the length of ξ' with respect to the Riemannian metric of Γ induced by the natural metric of $\mathbf{R}^n$.

Proof. (i) By the divergence theorem, it follows that for all $u \in C^2(\overline{\Omega})$

$$\begin{aligned} -\operatorname{Re} \iint_\Omega Au \cdot \overline{u}\, dx = {} & \sum_{i,j=1}^n \iint_\Omega a^{ij} \frac{\partial u}{\partial x_i} \cdot \overline{\frac{\partial u}{\partial x_j}}\, dx - \operatorname{Re} \int_\Gamma \frac{\partial u}{\partial \boldsymbol{\nu}} \cdot \overline{u}\, d\sigma \\ & + \iint_\Omega \left(\frac{1}{2} \sum_{i=1}^n \frac{\partial b^i}{\partial x_i} - c \right) |u|^2\, dx \\ & - \frac{1}{2} \int_\Gamma \sum_{i=1}^n b^i \cdot n_i\, |u|^2\, d\sigma. \end{aligned}$$

Hence, taking $u = P\varphi$ with $\varphi \in C^\infty(\Gamma)$, we obtain that

$$\begin{aligned} \operatorname{Re} \int_\Gamma \Pi\varphi \cdot \overline{\varphi}\, d\sigma = {} & \sum_{i,j=1}^n \iint_\Omega a^{ij} \frac{\partial (P\varphi)}{\partial x_i} \cdot \overline{\frac{\partial (P\varphi)}{\partial x_j}}\, dx \\ & + \iint_\Omega \left(\frac{1}{2} \sum_{i=1}^n \frac{\partial b^i}{\partial x_i} - c \right) |P\varphi|^2\, dx \\ & - \frac{1}{2} \int_\Gamma \sum_{i=1}^n b^i \cdot n_i\, |\varphi|^2\, d\sigma \\ \geq {} & a_0 \sum_{i=1}^n \iint_\Omega \left| \frac{\partial (P\varphi)}{\partial x_i} \right|^2 dx \\ & - C_1 \iint_\Omega |P\varphi|^2\, dx - C_2 \int_\Gamma |\varphi|^2\, d\sigma \\ \geq {} & a_0 \|P\varphi\|^2_{1,2} - (a_0 + C_1) \|P\varphi\|^2_{0,2} - C_2 \int_\Gamma |\varphi|^2\, d\sigma. \end{aligned} \tag{4.14}$$

But it is known (cf. [Mi, Theorem 3.16]) that, for each $\varepsilon > 0$, there exists a constant $C_\varepsilon > 0$ such that

$$\int_\Gamma |\varphi|^2 \, d\sigma \leq \varepsilon \|P\varphi\|_{1,2}^2 + C_\varepsilon \|P\varphi\|_{0,2}^2.$$

Hence, carrying this interpolation inequality into inequality (4.14) with $\varepsilon = a_0/2$, we have

$$\text{Re} \int_\Gamma \Pi\varphi \cdot \overline{\varphi} \, d\sigma \geq \frac{a_0}{2} \|P\varphi\|_{1,2}^2 - C_3 \|P\varphi\|_{0,2}^2. \tag{4.15}$$

Therefore, the desired inequality (4.12) follows from inequality (4.15), since the operator P maps $B^{\sigma-1/2,2}(\Gamma)$ isomorphically onto $N(A, \sigma, 2)$ for all $\sigma \in \mathbf{R}$.

(iii) It is known (cf. [Ho], [Ku], [Ty]) that the strong ellipticity (4.12) implies inequality (4.13) for the principal symbol of the operator Π.

CHAPTER V

PROOF OF THEOREM 1

In this chapter we consider the boundary value problem

$$(*)\qquad \begin{cases} Au = f & \text{in } \Omega, \\ Bu = a\frac{\partial u}{\partial \nu} + bu\big|_{\Gamma} = \varphi & \text{on } \Gamma, \end{cases}$$

and prove Theorem 1. We associate with problem $(*)$ a linear operator

$$\mathcal{A} = (A, B) : H^{s,p}(\Omega) \longmapsto H^{s-2,p}(\Omega) \times B_*^{s-1-1/p,p}(\Gamma).$$

In order to prove Theorem 1, it suffices to show that the operator $\mathcal{A}$ is *bijective*. Indeed, the continuity of the inverse of $\mathcal{A}$ follows immediately from an application of Banach's closed graph theorem, since $\mathcal{A}$ is a continuous operator.

Theorem 1 will be proved in a series of theorems (Theorems 5.1, 5.5 and 5.7) in the subsequent sections.

5.1 Regularity Theorem for Problem $(*)$

The next theorem states that the operator $\mathcal{A}$ has the *regularity property*:

Theorem 5.1. *Assume that conditions (H.1) and (H.2) are satisfied:*
(H.1) $a(x') \geq 0$ *and* $b(x') \geq 0$ *on* Γ.
(H.2) $b(x') > 0$ *on* $\Gamma_0 = \{x' \in \Gamma : a(x') = 0\}$.
Then we have for all $s \in \mathbf{R}$ *and* $1 < p < \infty$

$$u \in L^p(\Omega),\ Au \in H^{s-2,p}(\Omega),\ Bu \in B_*^{s-1-1/p,p}(\Gamma) \Longrightarrow u \in H^{s,p}(\Omega).$$

Proof. (1) By Theorem 4.11, we know that the question of regularity for solutions of problem $(*)$ is reduced to the corresponding question for the operator T, where

$$T = a\Pi + b,$$

and

$$\Pi\varphi = \frac{\partial}{\partial \nu}(P\varphi)\bigg|_{\Gamma}.$$

Typeset by $\mathcal{A}_{\mathcal{M}}\mathcal{S}$-TEX

The operator Π is a classical pseudo-differential operator of first order on the boundary Γ, and its complete symbol is given by the following (cf. [Ta1, Section 10.2]):

$$\left(p_1(x',\xi') + \sqrt{-1}\, q_1(x',\xi')\right) + \left(p_0(x',\xi') + \sqrt{-1}\, q_0(x',\xi')\right)$$
$$+ \text{ terms of order } \leq -1,$$

where (cf. (4.13))

$$p_1(x',\xi') \geq c_0|\xi'| \quad \text{on } T^*(\Gamma)\setminus\{0\}. \tag{5.1}$$

For example, if A is the usual Laplacian $\Delta = \partial^2/\partial x_1^2 + \cdots + \partial^2/\partial x_n^2$, then we have

$$p_1(x',\xi') = |\xi'|.$$

Thus, we obtain that the operator $T = a\Pi + b$ is a classical pseudo-differential operator of first order on the boundary Γ, and its complete symbol $t(x',\xi')$ is given by the following:

$$\begin{aligned} t(x',\xi') = {} & a(x')\left(p_1(x',\xi') + \sqrt{-1}\, q_1(x',\xi')\right) \\ & + \left([b(x') + a(x')p_0(x',\xi)] + \sqrt{-1}\, a(x')q_0(x',\xi)\right) \\ & + \text{terms of order} \leq -1. \end{aligned} \tag{5.2}$$

(2) Therefore, in order to prove Theorem 5.1, it suffices to show the following:

Lemma 5.2. *Assume that conditions (H.1) and (H.2) are satisfied. Then we have for all* $s \in \mathbf{R}$

$$\varphi \in \mathcal{D}'(\Gamma),\ T\varphi \in B^{s,p}(\Gamma) \implies \varphi \in B^{s,p}(\Gamma). \tag{5.3}$$

Furthermore, for any $t < s$, *there exists a constant* $C_{s,t} > 0$ *such that*

$$|\varphi|_{s,p} \leq C_{s,t}\left(|T\varphi|_{s,p} + |\varphi|_{t,p}\right). \tag{5.4}$$

Proof. (a) The proof of Lemma 5.2 is based on the following lemma:

Lemma 5.3. *If conditions (H.1) and (H.2) are satisfied, then, for each point* x' *of* Γ, *one can find a neighborhood* $U(x')$ *of* x' *such that:*

For any compact $K \subset U(x')$ *and any multi-indices* α, β, *there exist constants* $C_{K,\alpha,\beta} > 0$ *and* $C_K > 0$ *such that we have for all* $x' \in K$ *and* $|\xi'| \geq C_K$

$$\left|D_{\xi'}^{\alpha} D_{x'}^{\beta} t(x',\xi')\right| \leq C_{K,\alpha,\beta}\, |t(x',\xi')|\, (1+|\xi|)^{-|\alpha|+(1/2)|\beta|}, \tag{5.5a}$$

(5.5b) $$|t(x',\xi')|^{-1} \leq C_K.$$

Granting Lemma 5.3 for the moment, we shall prove Lemma 5.2.

(b) First we cover the boundary Γ by a finite number of local charts $\{(U_j,\chi_j)\}_{j=1}^m$ in each of which inequalities (5.5a) and (5.5b) hold. Since the operator T satisfies conditions (3.7a) and (3.7b) of Theorem 3.19 with $\mu = 0$, $\rho = 1$ and $\delta = 1/2$, it follows from an application of the same theorem that there exists a parametrix S in the class $L^0_{1,1/2}(U_j)$ for T. Let $\{\varphi_j\}_{j=1}^m$ be a partition of unity subordinate to the covering $\{U_j\}_{j=1}^m$, and choose a function $\psi_j \in C_0^\infty(U_j)$ such that $\psi_j = 1$ on $\operatorname{supp}\varphi_j$, so that $\varphi_j\psi_j = \varphi_j$.

Now one may assume that $\varphi \in B^{t,p}(\Gamma)$ for some $t < s$ and that $T\varphi \in B^{s,p}(\Gamma)$. We remark that the operator T can be written in the following form:

$$T = \sum_{j=1}^m \varphi_j T\psi_j + \sum_{j=1}^m \varphi_j T(1-\psi_j).$$

But the second terms $\varphi_j T(1-\psi_j)$ are in $L^{-\infty}(\Gamma)$, since $\varphi_j(1-\psi_j) = 0$. Hence we are reduced to the study of the first terms $\varphi_j T\psi_j$. This implies that we have only to prove the following *local* version of assertions (5.3) and (5.4):

(5.3′) $$\psi_j\varphi \in B^{t,p}(U_j),\ T\psi_j\varphi \in B^{s,p}(U_j) \implies \psi_j\varphi \in B^{s,p}(U_j);$$

(5.4′) $$|\psi_j\varphi|_{s,p} \leq C'_{s,t}\left(|T\psi_j\varphi|^2_{s,p} + |\psi_j\varphi|^2_{t,p}\right).$$

But, applying Theorem 3.17 to our situation, we obtain that the parametrix S maps $B^{\sigma,p}_{loc}(U_j)$ continuously into itself for all $\sigma \in \mathbf{R}$. This proves assertions (5.3′) and (5.4′), since we have $ST \equiv I \mod L^{-\infty}(U_j)$.

Lemma 5.2 is proved, apart from the proof of Lemma 5.3. □

(c) *Proof of Lemma 5.3*

(c-1) First we verify condition (5.5b):

By assertions (5.2) and (5.1), it follows that we have for $|\xi'|$ large enough

$$\begin{aligned}|t(x',\xi')| &\geq a(x')\,|p_1(x',\xi') + p_0(x',\xi')| + b(x') \\ &\geq \begin{cases} \frac{c_0}{2}\,a(x')|\xi'| + b(x') & \text{if } a(x') > 0, \\ b(x') & \text{if } a(x') = 0, \end{cases}\end{aligned}$$

so that

(5.6) $$|t(x',\xi')| \geq C\left(a(x')|\xi'| + 1\right),$$

since $b(x') > 0$ on $\Gamma_0 = \{x' \in \Gamma : a(x') = 0\}$. Here and in the following the letter C denotes a generic positive constant.

Inequality (5.6) implies condition (5.5b):

$$|t(x', \xi')| \geq C. \tag{5.7}$$

(c-2) Next we verify condition (5.5a) for $|\alpha| = 1$ and $|\beta| = 0$:
Since we have for $|\xi'|$ large enough

$$\left|D_{\xi'}^{\alpha} t(x', \xi')\right| \leq C \left(a(x') + |\xi'|^{-1}\right),$$

it follows from inequality (5.6) that

$$\begin{aligned}\left|D_{\xi'}^{\alpha} t(x', \xi')\right| &\leq C(1 + |\xi'|)^{-1} \left(a(x')|\xi'| + 1\right) \\ &\leq C(1 + |\xi'|)^{-1} \left|t(x', \xi')\right|.\end{aligned}$$

This inequality proves condition (5.5a) for $|\alpha| = 1$ and $|\beta| = 0$.

(c-3) We verify condition (5.5a) for $|\beta| = 1$ and $|\alpha| = 0$: To do so, we need the following elementary lemma on non-negative functions.

Lemma 5.4. *Let f be a non-negative C^2 function on $\mathbf{R}$ such that for some constant $c > 0$*

$$\sup_{x \in \mathbf{R}} |f''(x)| \leq c. \tag{5.8}$$

Then we have

$$|f'(x)| \leq \sqrt{2c}\,\sqrt{f(x)} \quad \text{on } \mathbf{R}. \tag{5.9}$$

Proof. In view of Taylor's formula, it follows that

$$0 \leq f(y) = f(x) + f'(x)(y - x) + \frac{f''(\xi)}{2} (y - x)^2,$$

where ξ is between x and y. Thus, letting $z = x - y$, we obtain from estimate (5.8) that

$$\begin{aligned}0 &\leq f(x) + f'(x)z + \frac{f''(\xi)}{2} z^2 \\ &\leq f(x) + f'(x)z + \frac{c}{2} z^2 \quad \text{for all } z \in \mathbf{R}.\end{aligned}$$

This implies inequality (5.9). □

Since we have for $|\xi'|$ large enough

$$\left|D_{x'}^{\beta} t(x', \xi')\right| \leq C \left(\left|D_{x'}^{\beta} a(x')\right| \cdot |\xi'| + a(x')|\xi'| + 1\right),$$

it follows from an application of Lemma 5.4 and inequalities (5.6) and (5.7) that

$$\begin{aligned}\left|D_{x'}^{\beta} t(x',\xi')\right| &\leq C\left[\left(\sqrt{a(x')}\,|\xi'|+1\right)+(a(x')|\xi'|+1)\right]\\ &\leq C\left[|\xi'|^{1/2}\left(a(x')|\xi'|+1\right)^{1/2}+(a(x')|\xi'|+1)\right]\\ &\leq C\,|t(x',\xi')|\left(|\xi'|^{1/2}\,|t(x',\xi')|^{-1/2}+1\right)\\ &\leq C\,|t(x',\xi')|\,(1+|\xi'|)^{1/2}.\end{aligned}$$

This inequality proves condition (5.5a) for $|\beta|=1$ and $|\alpha|=0$.

(c-4) Similarly, we can verify condition (5.5a) for the general case: $|\alpha|+|\beta|=k$, $k\in\mathbf{N}$.

Now the proof of Lemma 5.2, and hence that of Theorem 5.1, is complete. □

5.2 Uniqueness Theorem for Problem (∗)

The next theorem asserts that the operator $\mathcal{A}$ is *injective*:

Theorem 5.5. *Let* $1<p<\infty$. *Assume that conditions (H.1) and (H.2) are satisfied:*

(H.1) $a(x')\geq 0$ *and* $b(x')\geq 0$ *on* Γ.
(H.2) $b(x')>0$ *on* $\Gamma_0=\{x'\in\Gamma : a(x')=0\}$.

Then we have

$$u\in H^{2,p}(\Omega),\ Au=0 \text{ in } \Omega,\ Bu=0 \text{ on } \Gamma \implies u=0 \text{ in } \Omega. \tag{5.10}$$

Proof. In view of Theorem 5.1, it follows that

$$u\in H^{2,p}(\Omega),\ Au=0 \text{ in } \Omega,\ Bu=0 \text{ on } \Gamma \implies u\in C^{\infty}(\overline{\Omega}).$$

Therefore, the uniqueness result (5.10) is an immediate consequence of the following *maximum principle*:

Proposition 5.6. *Assume that conditions (H.1) and (H.2) are satisfied. Then we have*

$$v\in C^{2}(\overline{\Omega}),\ Av\geq 0 \text{ in } \Omega,\ Bv\leq 0 \text{ on } \Gamma \implies v\leq 0 \text{ on } \overline{\Omega}.$$

Proof. If v is a constant m, then we have

$$0\leq Av=mc \quad \text{in } \Omega.$$

This implies that m is non-positive, since $c\leq 0$ and $c\not\equiv 0$ in Ω.

Now we consider the case when v is not a constant. Assume to the contrary that

$$m = \max_{\overline{\Omega}} v > 0.$$

Then, applying the strong maximum principle (cf. Appendix, Theorem A.1) to the operator A, we obtain that there exists a point x_0' of Γ such that

$$\begin{cases} v(x_0') = m, \\ v(x) < v(x_0') \quad \text{for all } x \in \Omega. \end{cases}$$

Furthermore, it follows from an application of the boundary point lemma (cf. Appendix, Theorem A.2) that

$$\frac{\partial v}{\partial \boldsymbol{\nu}}(x_0') > 0.$$

Since we have

$$Bv(x_0') = a(x_0')\frac{\partial v}{\partial \boldsymbol{\nu}}(x_0') + b(x_0')v(x_0') \leq 0,$$

it follows from condition (H.1) that

$$a(x_0') = 0, \quad b(x_0') = 0.$$

This contradicts condition (H.2). □

5.3. Existence Theorem for Problem (*)

The next theorem asserts that the operator $\mathcal{A}$ is *surjective*:

Theorem 5.7. *Let $s > 1 + 1/p$ where $1 < p < \infty$. Assume that conditions (H.1) and (H.2) are satisfied:*

(H.1) $a(x') \geq 0$ and $b(x') \geq 0$ on Γ.
(H.2) $b(x') > 0$ on $\Gamma_0 = \{x' \in \Gamma : a(x') = 0\}$.

Then, for any $f \in H^{s-2,p}(\Omega)$ and any $\varphi \in B_*^{s-1-1/p,p}(\Gamma)$, the boundary value problem

$$(*) \qquad \begin{cases} Av = f & \text{in } \Omega, \\ a\frac{\partial v}{\partial \nu} + bv\big|_\Gamma = \varphi & \text{on } \Gamma \end{cases}$$

has a solution u in the space $H^{s,p}(\Omega)$.

5.3A Proof of Theorem 5.7. First, by Theorem 4.10, we know that

$$\operatorname{ind} \mathcal{A} = \operatorname{ind} \mathcal{T},$$

where the operator

$$\mathcal{T} : B^{s-1/p,p}(\Gamma) \longrightarrow B^{s-1/p,p}(\Gamma)$$

is defined as follows:

(a) The domain $D(\mathcal{T})$ of $\mathcal{T}$ is the space

$$D(\mathcal{T}) = \left\{ \varphi \in B^{s-1/p,p}(\Gamma) : T\varphi \in B^{s-1/p,p}(\Gamma) \right\}.$$

(b) $\mathcal{T}\varphi = T\varphi$, $\varphi \in D(\mathcal{T})$.

But Theorem 5.5 tells us that the operator $\mathcal{A}$ (or equivalently the operator $\mathcal{T}$) is injective. Hence, in order to prove the surjectivity, it suffices to show the following:

Proposition 5.8. *The index of the operator $\mathcal{T}$ is equal to* zero.

Proof. (1) We replace the operator A by the operator $A - \lambda$ with $\lambda \geq 0$, and consider instead of problem $(*)$ the following boundary value problem:

$$(*)_\lambda \qquad \begin{cases} (A - \lambda)u = f & \text{in } \Omega, \\ Bu = a\frac{\partial u}{\partial \nu} + bu\big|_\Gamma = \varphi & \text{on } \Gamma. \end{cases}$$

We associate with problem $(*)_\lambda$ a linear operator

$$\mathcal{A}(\lambda) = (A - \lambda, B) : H^{s,p}(\Omega) \longmapsto H^{s-2,p}(\Omega) \times B_*^{s-1-1/p,p}(\Gamma).$$

We remark that the operator $\mathcal{A}(\lambda)$ coincides with the operator $\mathcal{A}$ when $\lambda = 0$.

We reduce the study of problem $(*)_\lambda$ to that of a pseudo-differential operator on the boundary, just as in the proof of Theorem 5.1.

We can prove that Theorem 4.1 remains valid for the operator $A - \lambda$. That is, we have the following:

(a) The Dirichlet problem

$$\begin{cases} (A - \lambda)w = 0 & \text{in } \Omega, \\ w|_\Gamma = \varphi & \text{on } \Gamma \end{cases}$$

has a unique solution w in $H^{t,p}(\Omega)$ for any $\varphi \in B^{t-1/p,p}(\Gamma)$, where $t \in \mathbf{R}$.

(b) The Poisson operator

$$P(\lambda) : B^{t-1/p,p}(\Gamma) \longrightarrow H^{t,p}(\Omega),$$

defined by $w = P(\lambda)\varphi$, is an isomorphism of the space $B^{t-1/p,p}(\Gamma)$ onto the space $N(A-\lambda, t, p) = \{u \in H^{t,p}(\Omega) : (A-\lambda)u = 0 \text{ in } \Omega\}$ for all $t \in \mathbf{R}$; and its inverse is the trace operator on Γ.

Let $T(\lambda)$ be a classical pseudo-differential operator of first order on the boundary Γ defined as follows:

$$T(\lambda) = BP(\lambda) = a\Pi(\lambda) + b, \quad \lambda \geq 0,$$

where

$$\begin{aligned} \Pi(\lambda) : C^\infty(\Gamma) &\longrightarrow C^\infty(\Gamma) \\ \varphi &\longmapsto BP(\lambda)\varphi. \end{aligned}$$

Since the operator $T(\lambda) : C^\infty(\Gamma) \longrightarrow C^\infty(\Gamma)$ extends to a continuous linear operator $T(\lambda) : B^{t,p}(\Gamma) \longrightarrow B^{t-1,p}(\Gamma)$ for all $t \in \mathbf{R}$, one can introduce a densely defined, closed linear operator

$$\mathcal{T}(\lambda) : B^{s-1/p,p}(\Gamma) \longrightarrow B^{s-1/p,p}(\Gamma)$$

as follows.

(α) The domain $D(\mathcal{T}(\lambda))$ of $\mathcal{T}(\lambda)$ is the space

$$D(\mathcal{T}(\lambda)) = \left\{\varphi \in B^{s-1/p,p}(\Gamma) : T(\lambda)\varphi \in B^{s-1/p,p}(\Gamma)\right\}.$$

(β) $\mathcal{T}(\lambda)\varphi = T(\lambda)\varphi$, $\varphi \in D(\mathcal{T}(\lambda))$.

We remark that the operator $\mathcal{T}(\lambda)$ coincides with the operator $\mathcal{T}$ when $\lambda = 0$.

Then we can obtain the following results (cf. Theorem 4.10):

(I) The null space $N(\mathcal{A}(\lambda))$ of $\mathcal{A}(\lambda)$ has finite dimension if and only if the null space $N(\mathcal{T}(\lambda))$ of $\mathcal{T}(\lambda)$ has finite dimension, and we have

$$\dim N(\mathcal{A}(\lambda)) = \dim N(\mathcal{T}(\lambda)).$$

(II) The range $R(\mathcal{A}(\lambda))$ of $\mathcal{A}(\lambda)$ is closed if and only if the range $R(\mathcal{T}(\lambda))$ of $\mathcal{T}(\lambda)$ is closed; and $R(\mathcal{A}(\lambda))$ has finite codimension if and only if $R(\mathcal{T}(\lambda))$ has finite codimension, and we have

$$\operatorname{codim} R(\mathcal{A}(\lambda)) = \operatorname{codim} R(\mathcal{T}(\lambda)).$$

(III) The operator $\mathcal{A}(\lambda)$ is a Fredholm operator if and only if the operator $\mathcal{T}(\lambda)$ is a Fredholm operator, and we have

$$\operatorname{ind} \mathcal{A}(\lambda) = \operatorname{ind} \mathcal{T}(\lambda).$$

(2) To study problem $(*)_\lambda$, we shall make use of a method essentially due to Agmon (cf. [Ag], [LM] and also [Ta2, Section 8.4]). This is a technique of

treating a spectral parameter λ as a second-order differential operator of an extra variable and relating the old problem to a new one with the additional variable.

We introduce an auxiliary variable y of the unit circle

$$S = \mathbf{R}/2\pi\mathbf{Z},$$

and replace the parameter $-\lambda$ by the second-order differential operator

$$\frac{\partial^2}{\partial y^2}.$$

That is, we replace the operator $A - \lambda$ by the operator

$$A + \frac{\partial^2}{\partial y^2},$$

and consider instead of problem $(*)_\lambda$ the following boundary value problem:

$$(\tilde{*}) \qquad \begin{cases} \widetilde{A}\tilde{u} := \left(A + \frac{\partial^2}{\partial y^2}\right)\tilde{u} = \tilde{f} & \text{in } \Omega \times S, \\ B\tilde{u} = a\frac{\partial \tilde{u}}{\partial \nu} + b\tilde{u}\big|_{\Gamma \times S} = \tilde{\varphi} & \text{on } \Gamma \times S. \end{cases}$$

We can prove that Theorem 4.1 remains valid for the operator $\widetilde{A} = A + \partial^2/\partial y^2$.

($\tilde{a}$) The Dirichlet problem

$$\begin{cases} \widetilde{A}\tilde{w} = 0 & \text{in } \Omega \times S, \\ \tilde{w}|_{\Gamma \times S} = \tilde{\varphi} & \text{on } \Gamma \times S \end{cases}$$

has a unique solution $\tilde{w}$ in $H^{t,p}(\Omega \times S)$ for any $\tilde{\varphi} \in B^{t-1/p,p}(\Gamma \times S)$, where $t \in \mathbf{R}$.

($\tilde{b}$) The Poisson operator

$$\widetilde{P} : B^{t-1/p,p}(\Gamma \times S) \longrightarrow H^{t,p}(\Omega \times S),$$

defined by $\tilde{w} = \widetilde{P}\tilde{\varphi}$, is an isomorphism of the space $B^{t-1/p,p}(\Gamma \times S)$ onto the space $N(\widetilde{A}, t, p) = \{\tilde{u} \in H^{t,p}(\Omega \times S) : \widetilde{A}\tilde{u} = 0 \text{ in } \Omega \times S\}$ for all $t \in \mathbf{R}$; and its inverse is the trace operator on $\Gamma \times S$.

We let

$$\begin{aligned} \widetilde{T} : C^\infty(\Gamma \times S) &\longrightarrow C^\infty(\Gamma \times S) \\ \tilde{\varphi} &\longmapsto B\widetilde{P}\tilde{\varphi}. \end{aligned}$$

Then the operator $\widetilde{T}$ can be decomposed as follows:

$$\widetilde{T} = a\widetilde{\Pi} + b,$$

where

$$\widetilde{\Pi}\tilde{\varphi} = \frac{\partial}{\partial \boldsymbol{\nu}} \left(\widetilde{P}\tilde{\varphi}\right)\Big|_{\Gamma \times S}.$$

The operator $\widetilde{\Pi}$ is a classical pseudo-differential operator of first order on the boundary $\Gamma \times S$, and its complete symbol is given by the following:

$$\begin{aligned}&\left[\tilde{p}_1(x', \xi', y, \eta) + \sqrt{-1}\, \tilde{q}_1(x', \xi', y, \eta)\right] \\ &+ \left[\tilde{p}_0(x', \xi', y, \eta) + \sqrt{-1}\, \tilde{q}_0(x', \xi', y, \eta)\right] + \text{terms of order} \leq -1,\end{aligned}$$

where (cf. (4.13))

$$\tilde{p}_1(x', \xi', y, \eta) \geq \tilde{c}_0 \sqrt{|\xi'|^2 + \eta^2} \quad \text{on } T^*(\Gamma \times S) \setminus \{0\}. \tag{5.11}$$

For example, if A is the usual Laplacian $\Delta = \partial^2/\partial x_1^2 + \cdots + \partial^2/\partial x_n^2$, then we have

$$\tilde{p}_1(x', \xi', y, \eta) = \sqrt{|\xi'|^2 + \eta^2}.$$

Thus we find that the operator

$$\widetilde{T} = a\widetilde{\Pi} + b$$

is a classical pseudo-differential operator of first order on the boundary $\Gamma \times S$ and its complete symbol is given by the following:

$$\begin{aligned}&a(x')\left[\tilde{p}_1(x', \xi', y, \eta) + \sqrt{-1}\, \tilde{q}_1(x', \xi', y, \eta)\right] \\ &+ \left[(b(x') + a(x')\tilde{p}_0(x', \xi', y, \eta)) + \sqrt{-1}\, a(x')\tilde{q}_0(x', \xi', y, \eta)\right] \\ &+ \text{terms of order} \leq -1.\end{aligned} \tag{5.12}$$

Then, by virtue of assertions (5.12) and (5.11), it is easy to verify that the operator $\widetilde{T}$ satisfies conditions (3.7a) and (3.7b) of Theorem 3.19 with $\mu = 0$, $\rho = 1$ and $\delta = 1/2$, just as in the proof of Lemma 5.2. Hence there exists a parametrix $\widetilde{S}$ in the class $L^0_{1,1/2}(\Gamma \times S)$ for the operator $\widetilde{T}$.

Therefore we obtain the following result, analogous to Lemma 5.2:

Lemma 5.9. *Assume that conditions (H.1) and (H.2) are satisfied. Then we have for all* $s \in \mathbf{R}$

$$\tilde{\varphi} \in \mathcal{D}'(\Gamma \times S),\ \widetilde{T}\tilde{\varphi} \in B^{s,p}(\Gamma \times S) \implies \tilde{\varphi} \in B^{s,p}(\Gamma \times S).$$

Furthermore, for any $t < s$, *there exists a constant* $\widetilde{C}_{s,t} > 0$ *such that*

$$|\tilde{\varphi}|_{s,p} \leq \widetilde{C}_{s,t} \left(|\widetilde{T}\tilde{\varphi}|_{s,p} + |\tilde{\varphi}|_{t,p}\right). \tag{5.13}$$

We introduce a densely defined, closed linear operator

$$\widetilde{T} : B^{s-1/p,p}(\Gamma \times S) \longrightarrow B^{s-1/p,p}(\Gamma \times S)$$

as follows.

($\tilde{\alpha}$) The domain $D(\widetilde{T})$ of $\widetilde{T}$ is the space

$$D(\widetilde{T}) = \left\{\tilde{\varphi} \in B^{s-1/p,p}(\Gamma \times S) : \widetilde{T}\tilde{\varphi} \in B^{s-1/p,p}(\Gamma \times S)\right\}.$$

($\tilde{\beta}$) $\widetilde{T}\tilde{\varphi} = \widetilde{T}\tilde{\varphi}$, $\tilde{\varphi} \in D(\widetilde{T})$.

Then the most fundamental relationship between the operators $\widetilde{T}$ and $T(\lambda)$ ($\lambda \geq 0$) is the following:

Proposition 5.10. *If* $\operatorname{ind}\widetilde{T}$ *is finite, then there exists a finite subset* K *of* $\mathbf{Z}$ *such that the operator* $T(\lambda')$ *is bijective for all* $\lambda' = \ell^2$ *satisfying* $\ell \in \mathbf{Z} \setminus K$.

Granting Proposition 5.10 for the moment, we shall prove Theorem 5.7.

(3) **End of Proof of Theorem 5.7**

(3-1) We show that if conditions (H.1) and (H.2) are satisfied, then we have

$$\operatorname{ind}\widetilde{T} < \infty. \tag{5.14}$$

To this end, we need a useful criterion for Fredholm operators (cf. [Ta2, Theorem 3.7.6]):

Lemma 5.11 (Peetre). *Let* X, Y, Z *be Banach spaces such that* $X \subset Z$ *is a compact injection, and let* T *be a closed linear operator from* X *into* Y *with domain* $D(T)$*. Then the following two conditions are equivalent:*

(i) The null space $N(T)$ *of* T *has finite dimension and the range* $R(T)$ *of* T *is closed in* Y*.*

(ii) There is a constant $C > 0$ *such that*

$$\|x\|_X \leq C\left(\|Tx\|_Y + \|x\|_Z\right), \quad x \in D(T).$$

Now, estimate (5.13) gives that

$$|\tilde{\varphi}|_{s-1/p,p} \leq \widetilde{C}_{s,t}\left(|\widetilde{T}\tilde{\varphi}|_{s-1/p,p} + |\tilde{\varphi}|_{t,p}\right), \quad \tilde{\varphi} \in D(\widetilde{T}), \tag{5.15}$$

where $t < s - 1/p$. But it follows from an application of Rellich's theorem that the injection $B^{s-1/p,p}(\Gamma \times S) \longrightarrow B^{t,p}(\Gamma \times S)$ is *compact* (or completely continuous) for $t < s - 1/p$. Thus, applying Lemma 5.11 with

$$X = Y = B^{s-1/p,p}(\Gamma \times S),$$

$$Z = B^{t,p}(\Gamma \times S),$$
$$T = \widetilde{T},$$

we obtain that the range $R(\widetilde{T})$ is closed in $B^{s-1/p,p}(\Gamma \times S)$ and

$$\dim N(\widetilde{T}) < \infty. \tag{5.16}$$

On the other hand, by formula (5.12), we find that the symbol of the adjoint $\widetilde{T}^*$ is given by the following (cf. Theorem 3.12):

$$\begin{aligned}
&a(x')\left(\tilde{p}_1(x',\xi',y,\eta) - \sqrt{-1}\,\tilde{q}_1(x',\xi',y,\eta)\right) \\
&+ \left(\left[b(x') + a(x')\tilde{p}_0(x',\xi',y,\eta) - \sum_{j=1}^{2} \partial_{x_j}\left(a(x')\cdot \partial_{\xi_j}\tilde{q}_1(x',\xi',y,\eta)\right)\right]\right. \\
&\left. - \sqrt{-1}\left[a(x')\tilde{q}_0(x',\xi',y,\eta) + \sum_{j=1}^{2} \partial_{x_j}\left(a(x')\cdot \partial_{\xi_j}\tilde{p}_1(x',\xi',y,\eta)\right)\right]\right) \\
&+ \text{terms of order} \le -1.
\end{aligned}$$

But, by virtue of Lemma 5.4, it follows that

$$\partial_{x_j} a(x') = 0 \text{ on } \Gamma_0 = \{x' \in \Gamma : a(x') = 0\}.$$

Thus one can easily verify that the pseudo-differential operator $\widetilde{T}^*$ satisfies conditions (3.7a) and (3.7b) of Theorem 3.19 with $\mu = 0$, $\rho = 1$ and $\delta = 1/2$. This implies that estimate (5.15) remains valid for the adjoint operator $\widetilde{T}^*$ of $\widetilde{T}$:

$$|\tilde{\psi}|_{-s+1/p,p'} \le \widetilde{C}_{s,\tau}\left(|\widetilde{T}^*\tilde{\psi}|_{-s+1/p,p'} + |\tilde{\psi}|_{\tau,p'}\right), \quad \tilde{\psi} \in D(\widetilde{T}^*),$$

where $\tau < -s + 1/p$ and $p' = p/(p-1)$, the exponent conjugate to p. Hence we have by the closed range theorem (cf. [Yo, Chapter VII, Section 5]) and Lemma 5.11

$$\operatorname{codim} R(\widetilde{T}) = \dim N(\widetilde{T}^*) < \infty, \tag{5.17}$$

since the injection $B^{-s+1/p,p'}(\Gamma \times S) \longrightarrow B^{\tau,p'}(\Gamma \times S)$ is compact for $\tau < -s + 1/p$.

Therefore, assertion (5.14) follows from assertions (5.16) and (5.17).

(3-2) By assertion (5.14), we can apply Proposition 5.10 to obtain that the operator $T(\ell^2) : B^{s-1/p,p}(\Gamma) \longrightarrow B^{s-1/p,p}(\Gamma)$ is bijective if $\ell \in \mathbf{Z} \setminus K$ for some finite subset K of $\mathbf{Z}$. In particular, we have

$$\operatorname{ind} T(\lambda_0) = 0 \quad \text{if } \lambda_0 = \ell^2,\ \ell \in \mathbf{Z} \setminus K. \tag{5.18}$$

But it is easy to see that the symbol $t(x', \xi'; \lambda)$ of the operator

$$T(\lambda) = a\Pi(\lambda) + b, \quad \lambda \geq 0,$$

has the following asymptotic expansion:

$$\begin{aligned}(5.19) \qquad t(x', \xi'; \lambda) = {} & a(x') \left[p_1(x', \xi') + \sqrt{-1}\, q_1(x', \xi')\right] \\ & + \left[(b(x') + a(x')p_0(x', \xi)) + \sqrt{-1}\, a(x')q_0(x', \xi')\right] \\ & + \text{terms of order} \leq -1 \text{ depending on } \lambda.\end{aligned}$$

Thus we can find a classical pseudo-differential operator $K(\lambda_0)$ of order -1 on the boundary Γ such that

$$\mathcal{T} = T(\lambda_0) + K(\lambda_0).$$

Furthermore, Rellich's theorem tells us that the operator

$$K(\lambda_0) : B^{s-1/p,p}(\Gamma) \longrightarrow B^{s-1/p,p}(\Gamma)$$

is *compact*. Hence we have

$$(5.20) \qquad \operatorname{ind} \mathcal{T} = \operatorname{ind} \mathcal{T}(\lambda_0).$$

Therefore, Proposition 5.8 follows from assertions (5.18) and (5.20).

Theorem 5.7 is proved, apart from the proof of Proposition 5.10.

5.3B Proof of Proposition 5.10. (i) First we study the null spaces $N(\widetilde{\mathcal{T}})$ and $N(\mathcal{T}(\lambda'))$ when $\lambda' = \ell^2$ with $\ell \in \mathbf{Z}$:

$$\begin{aligned} N(\widetilde{\mathcal{T}}) &= \left\{\tilde{\varphi} \in B^{s-1/p,p}(\Gamma \times S) : \widetilde{\mathcal{T}}\tilde{\varphi} = 0\right\}, \\ N(\mathcal{T}(\lambda')) &= \left\{\varphi \in B^{s-1/p,p}(\Gamma) : T(\lambda')\varphi = 0\right\}. \end{aligned}$$

Since the pseudo-differential operators $\widetilde{T}$ and $T(\lambda')$ are both *hypoelliptic*, it follows that

$$\begin{aligned} N(\widetilde{\mathcal{T}}) &= \left\{\tilde{\varphi} \in C^\infty(\Gamma \times S) : \widetilde{T}\tilde{\varphi} = 0\right\}, \\ N(\mathcal{T}(\lambda')) &= \left\{\varphi \in C^\infty(\Gamma) : T(\lambda')\varphi = 0\right\}. \end{aligned}$$

Therefore, we can apply [Ta2, Proposition 8.4.6] to obtain the following most important relationship between the null spaces $N(\widetilde{\mathcal{T}})$ and $N(\mathcal{T}(\lambda'))$ when $\lambda' = \ell^2$ with $\ell \in \mathbf{Z}$:

Lemma 5.12. *The following two conditions are equivalent:*
(1) $\dim N(\widetilde{\mathcal{T}}) < \infty$.
(2) There exists a finite subset I of $\mathbf{Z}$ *such that*

$$\begin{cases} \dim N\left(\mathcal{T}(\ell^2)\right) < \infty & \text{if } \ell \in I, \\ \dim N\left(\mathcal{T}(\ell^2)\right) = 0 & \text{if } \ell \notin I. \end{cases}$$

Moreover, in this case, we have

$$N(\widetilde{\mathcal{T}}) = \bigoplus_{\ell \in I} N\left(\mathcal{T}(\ell^2)\right) \otimes \ell^{i\ell y},$$

$$\dim N(\widetilde{\mathcal{T}}) = \sum_{\ell \in I} \dim N(\mathcal{T}(\ell^2)).$$

(ii) Next we study the ranges $R(\widetilde{\mathcal{T}})$ and $R(\mathcal{T}(\lambda'))$ when $\lambda' = \ell^2$ with $\ell \in \mathbf{Z}$. To do so, we consider the adjoint operators $\widetilde{\mathcal{T}}^*$ and $\mathcal{T}(\lambda')^*$ of $\widetilde{\mathcal{T}}$ and $\mathcal{T}(\lambda')$, respectively.

The next lemma allows us to give a characterization of the adjoint operators $\widetilde{\mathcal{T}}^*$ and $\mathcal{T}(\lambda')^*$ in terms of pseudo-differential operators (cf. [Ta2, Lemma 8.4.8]):

Lemma 5.13. *Let M be a compact* C^∞ *manifold without boundary. If T is a classical pseudo-differential operator of order m on M, we define a densely defined, closed linear operator*

$$\mathcal{T} : B^{s,p}(M) \longrightarrow B^{s-m+1,p}(M) \quad (s \in \mathbf{R})$$

as follows.

(a) The domain $D(\mathcal{T})$ *of* $\mathcal{T}$ *is the space*

$$D(\mathcal{T}) = \left\{\varphi \in H^{s,p}(M) : T\varphi \in H^{s-m+1,p}(M)\right\}.$$

(b) $\mathcal{T}\varphi = T\varphi$, $\varphi \in D(\mathcal{T})$.

Then the adjoint operator $\mathcal{T}^*$ *of* $\mathcal{T}$ *is characterized as follows:*

(c) The domain $D(\mathcal{T}^*)$ *of* $\mathcal{T}^*$ *is contained in the space*

$$\left\{\psi \in B^{-s+m-1,p'}(M) : T^*\psi \in B^{-s,p'}(M)\right\},$$

where $p' = p/(p-1)$ *and* $T^* \in L^m_{\mathrm{cl}}(M)$ *is the adjoint of T.*

(d) $\mathcal{T}^*\psi = T^*\psi$, $\psi \in D(\mathcal{T}^*)$.

We remark that the pseudo-differential operators $T(\lambda)^*$ and $\widetilde{T}^*$ also satisfy conditions (3.7a) and (3.7b) of Theorem 3.19 with $\mu = 0$, $\rho = 1$ and $\delta = 1/2$; hence they are *hypoelliptic*.

Therefore, applying Lemma 5.13 to the operators $\widetilde{T}$ and $T(\lambda')$, we obtain the following:

Lemma 5.14. *The null spaces* $N(\widetilde{\mathcal{T}}^*)$ *and* $N(\mathcal{T}(\lambda')^*)$ *are characterized respectively as follows:*

$$N(\widetilde{\mathcal{T}}^*) = \left\{\tilde{\psi} \in C^\infty(\Gamma \times S) : \widetilde{T}^* \tilde{\psi} = 0\right\},$$
$$N(\mathcal{T}(\lambda')^*) = \{\psi \in C^\infty(\Gamma) : T(\lambda')^* \psi = 0\}.$$

By Lemma 5.14, we find that Lemma 5.12 remains valid for the adjoint operators $\widetilde{\mathcal{T}}^*$ and $\mathcal{T}(\lambda')^*$ (cf. [Ta2, Lemma 8.4.10]):

Lemma 5.15. *The following two conditions are equivalent:*
(1) $\dim N(\widetilde{\mathcal{T}}^*) < \infty$.
(2) There exists a finite subset J of $\mathbf{Z}$ such that

$$\begin{cases} \dim N\left(\mathcal{T}(\ell^2)^*\right) < \infty & \text{if } \ell \in J, \\ \dim N\left(\mathcal{T}(\ell^2)^*\right) = 0 & \text{if } \ell \notin J. \end{cases}$$

Moreover, in this case, we have

$$\dim N(\widetilde{\mathcal{T}}^*) = \sum_{\ell \in J} \dim N\left(\mathcal{T}(\ell^2)^*\right).$$

Hence, combining Lemma 5.15 and the closed range theorem, we obtain the most important relationship between $\operatorname{codim} R(\widetilde{\mathcal{T}})$ and $\operatorname{codim} R(\mathcal{T}(\lambda'))$ when $\lambda' = \ell^2$, $\ell \in \mathbf{Z}$ (cf. [Ta2, Proposition 8.4.11]):

Lemma 5.16. *The following two conditions are equivalent:*
(1) $\operatorname{codim} R(\widetilde{\mathcal{T}}) < \infty$.
(2) There exists a finite subset J of $\mathbf{Z}$ such that

$$\begin{cases} \operatorname{codim} R\left(\mathcal{T}(\ell^2)\right) < \infty & \text{if } \ell \in J, \\ \operatorname{codim} N\left(\mathcal{T}(\ell^2)\right) = 0 & \text{if } \ell \notin J. \end{cases}$$

Moreover, in this case, we have

$$\operatorname{codim} R(\widetilde{\mathcal{T}}) = \sum_{\ell \in J} \operatorname{codim} R\left(\mathcal{T}(\ell^2)\right).$$

(iii) Proposition 5.10 is an immediate consequence of Lemmas 5.12 and 5.16, with $K = I \cup J$.

Now the proof of Proposition 5.8, and hence that of Theorem 5.7, is complete. □

CHAPTER VI

PROOF OF THEOREM 2

In this chapter we prove Theorem 2. More precisely, we prove a generation theorem for analytic semigroups (Theorem 6.8) for the operator $\mathfrak{A}$ from $L^p(\Omega)$ into itself defined by the following:

(*a*) The domain of definition $D(\mathfrak{A})$ of $\mathfrak{A}$ is the set

$$D(\mathfrak{A}) = \left\{ u \in H^{2,p}(\Omega) : Bu = a\frac{\partial u}{\partial \nu} + bu\Big|_{\Gamma} = 0 \right\}.$$

(*b*) $\mathfrak{A}u = Au$, $u \in D(\mathfrak{A})$.

Once again Agmon's method plays an important role in the proof of the surjectivity of the operator $\mathfrak{A} - \lambda I$ (Proposition 6.7).

6.1 A Priori Estimates

In this section we study the operator $\mathfrak{A}$, and prove *a priori* estimates for the operator $\mathfrak{A} - \lambda I$ (Theorem 6.3) which will play a fundamental role in the next section. In the proof we make good use of Agmon's method (Proposition 6.4).

First, we have the following:

Lemma 6.1. *Assume that conditions (H.1) and (H.2) are satisfied:*
(H.1) $a(x') \geq 0$ *and* $b(x') \geq 0$ *on* Γ.
(H.2) $b(x') > 0$ *on* $\Gamma_0 = \{x' \in \Gamma : a(x') = 0\}$.
Then we have the a priori *estimate*

$$\|u\|_{2,p} \leq C\|Au\|_p, \quad u \in D(\mathfrak{A}). \tag{6.1}$$

Proof. Estimate (6.1) follows immediately from Theorem 1 with $s = 2$ and $\varphi = 0$. □

Corollary 6.2. *The operator $\mathfrak{A}$ is a closed operator.*

Proof. Let $\{u_j\}$ be an arbitrary sequence in the domain $D(\mathfrak{A})$ such that

$$\begin{cases} u_j \longrightarrow u & \text{in } L^p(\Omega), \\ Au_j \longrightarrow v & \text{in } L^p(\Omega). \end{cases}$$

Typeset by $\mathcal{A}_{\mathcal{M}}\mathcal{S}$-TEX

Then, applying estimate (6.1) to the sequence $\{u_j\}$, we find that $\{u_j\}$ is a Cauchy sequence in the space $H^{2,p}(\Omega)$, so that $u \in H^{2,p}(\Omega)$ and

$$u_j \longrightarrow u \quad \text{in } H^{2,p}(\Omega).$$

Hence we have

$$Au = \lim_{j\to\infty} Au_j = v \quad \text{in } L^p(\Omega),$$

and also by Proposition 4.8

$$Bu = \lim_{j\to\infty} Bu_j = 0 \quad \text{in } B_*^{1-1/p,p}(\Gamma).$$

This proves that $u \in D(\mathfrak{A})$ and $\mathfrak{A}u = v$. □

The next theorem is an essential step in the proof of Theorem 2:

Theorem 6.3. *Assume that conditions (H.1) and (H.2) are satisfied. Then, for every $-\pi < \theta < \pi$, there exists a constant $R(\theta) > 0$ depending on θ such that if $\lambda = r^2 e^{i\theta}$ and $|\lambda| = r^2 \geq R(\theta)$, we have for all $u \in H^{2,p}(\Omega)$ satisfying $Bu = 0$ on Γ (i.e., $u \in D(\mathfrak{A})$)*

$$|u|_{2,p} + |\lambda|^{1/2} \cdot |u|_{1,p} + |\lambda| \cdot \|u\|_p \leq C(\theta)\, \|(A-\lambda)u\|_p, \tag{6.2}$$

with a constant $C(\theta) > 0$ depending on θ. Here $|\cdot|_{j,p}$ ($j = 1, 2$) is the seminorm on the space $H^{2,p}(\Omega)$ defined by

$$|u|_{j,p} = \left(\int_\Omega \sum_{|\alpha|=j} |D^\alpha u(x)|^p \, dx \right)^{1/p}.$$

Proof. (1) We replace the operator $A - \lambda$ by the operator

$$A + e^{i\theta} \frac{\partial^2}{\partial y^2}, \qquad -\pi < \theta < \pi,$$

and consider instead of the problem

$$(*)_\lambda \qquad \begin{cases} (A-\lambda)u = f & \text{in } \Omega, \\ Bu = a\frac{\partial u}{\partial \nu} + bu\big|_\Gamma = 0 & \text{on } \Gamma \end{cases}$$

the following boundary value problem:

$$(\widetilde{*}) \qquad \begin{cases} \widetilde{A}(\theta)\tilde{u} := \left(A + e^{i\theta}\frac{\partial^2}{\partial y^2}\right)\tilde{u} = \tilde{f} & \text{in } \Omega \times S, \\ B\tilde{u} := a\frac{\partial \tilde{u}}{\partial \nu} + b\tilde{u}\big|_{\Gamma\times S} = 0 & \text{on } \Gamma \times S. \end{cases}$$

We remark that the operator $\widetilde{A}(\theta)$ is *elliptic* for $-\pi < \theta < \pi$.

Then we have the following result, analogous to Lemma 6.1:

Proposition 6.4. *Assume that conditions (H.1) and (H.2) are satisfied. Then we have for all $\tilde{u} \in H^{2,p}(\Omega \times S)$ satisfying $B\tilde{u} = 0$ on $\Gamma \times S$*

$$\|\tilde{u}\|_{2,p} \leq \widetilde{C}(\theta) \left(\left\| \widetilde{\Lambda}(\theta)\tilde{u} \right\|_p + \|\tilde{u}\|_p \right), \tag{6.3}$$

with a constant $\widetilde{C}(\theta) > 0$ depending on θ.

Proof. We reduce the study of problem ($\tilde{*}$) to that of a pseudo-differential operator on the boundary, just as in problem ($*$).

We can prove that Theorems 4.3 and 4.4 remain valid for the operator $\widetilde{\Lambda}(\theta) = A + e^{i\theta}\partial^2/\partial y^2$, $-\pi < \theta < \pi$:

($\tilde{a}$) The Dirichlet problem

$$\begin{cases} \widetilde{\Lambda}(\theta)\tilde{w} = 0 & \text{in } \Omega \times S, \\ \tilde{w}|_{\Gamma \times S} = \tilde{\varphi} & \text{on } \Gamma \times S \end{cases}$$

has a unique solution $\tilde{w}$ in $H^{t,p}(\Omega \times S)$ for any $\tilde{\varphi} \in B^{t-1/p,p}(\Gamma \times S)$, where $t \in \mathbf{R}$.

($\tilde{b}$) The Poisson operator

$$\widetilde{P}(\theta) : B^{t-1/p,p}(\Gamma \times S) \longrightarrow H^{t,p}(\Omega \times S),$$

defined by $\tilde{w} = \widetilde{P}(\theta)\tilde{\varphi}$, is an isomorphism of $B^{t-1/p,p}(\Gamma \times S)$ onto the space $N(\widetilde{\Lambda}(\theta), t, p) = \{\tilde{u} \in H^{t,p}(\Omega \times S) : \widetilde{\Lambda}(\theta)\tilde{u} = 0 \text{ in } \Omega \times S\}$ for all $t \in \mathbf{R}$; and its inverse is the trace operator on $\Gamma \times S$.

We let

$$\begin{aligned} \widetilde{T}(\theta) : C^\infty(\Gamma \times S) &\longrightarrow C^\infty(\Gamma \times S) \\ \tilde{\varphi} &\longmapsto B\widetilde{P}(\theta)\tilde{\varphi}. \end{aligned}$$

Then the operator $\widetilde{T}(\theta)$ can be decomposed as follows:

$$\widetilde{T}(\theta) = a\widetilde{\Pi}(\theta) + b$$

where

$$\widetilde{\Pi}(\theta)\tilde{\varphi} = \frac{\partial}{\partial \boldsymbol{\nu}} \left(\widetilde{P}(\theta)\tilde{\varphi} \right)\Big|_{\Gamma \times S}.$$

The operator $\widetilde{\Pi}(\theta)$ is a classical pseudo-differential operator of first order on the boundary $\Gamma \times S$, and its complete symbol is given by the following (cf. [Ta2, Section 10.2]):

$$\left(\tilde{p}_1(x', \xi', y, \eta; \theta) + \sqrt{-1}\, \tilde{q}_1(x', \xi', y, \eta; \theta) \right)$$

$$+ \left(\tilde{p}_0(x', \xi', y, \eta; \theta) + \sqrt{-1}\, \tilde{q}_0(x', \xi', y, \eta; \theta)\right) + \text{terms of order} \leq -1,$$

where (cf. (5.11))

$$\tilde{p}_1(x', \xi', y, \eta; \theta) \geq \tilde{c}_\theta \sqrt{|\xi'|^2 + \eta^2} \quad \text{on } T^*(\Gamma \times S) \setminus \{0\}. \tag{6.4}$$

For example, if A is the usual Laplacian $\Delta = \partial^2/\partial x_1^2 + \cdots + \partial^2/\partial x_n^2$, then we have

$$\tilde{p}_1(x', \xi', y, \eta; \theta) = \left[\frac{\left[\left(|\xi'|^2 + \cos\theta \cdot \eta^2\right)^2 + \sin^2\theta \cdot \eta^4 \right]^{1/2} + \left(|\xi'|^2 + \cos\theta \cdot \eta^2\right)}{2} \right]^{1/2}.$$

Therefore, the operator $\widetilde{T}(\theta) = a\widetilde{\Pi}(\theta) + b$ is a classical pseudo-differential operator of first order on the boundary $\Gamma \times S$ and its complete symbol is given by the following:

$$\begin{aligned} &a(x')\left(\tilde{p}_1(x', \xi', y, \eta; \theta) + \sqrt{-1}\, \tilde{q}_1(x', \xi', y, \eta; \theta)\right) \\ &+ \left([b(x') + a(x')\tilde{p}_0(x', \xi', y, \eta; \theta)] + \sqrt{-1}\, a(x')\tilde{q}_0(x', \xi', y, \eta; \theta)\right) \\ &+ \text{terms of order} \leq -1. \end{aligned} \tag{6.5}$$

Then, by virtue of assertions (6.5) and (6.4), one can verify that the operator $\widetilde{T}(\theta)$ satisfies conditions (3.7a) and (3.7b) of Theorem 3.19 with $\mu = 0$, $\rho = 1$ and $\delta = 1/2$, just as in the proof of Lemma 5.3. Hence we obtain the following result, analogous to Lemma 5.2:

Lemma 6.5. *Assume that conditions (H.1) and (H.2) are satisfied. Then we have for all* $s \in \mathbf{R}$

$$\tilde{\varphi} \in \mathcal{D}'(\Gamma \times S), \ \widetilde{T}(\theta)\tilde{\varphi} \in B^{s,p}(\Gamma \times S) \implies \tilde{\varphi} \in B^{s,p}(\Gamma \times S).$$

Furthermore, for any $t < s$, *there exists a constant* $\widetilde{C}_{s,t} > 0$ *such that*

$$|\tilde{\varphi}|_{s,p} \leq \widetilde{C}_{s,t} \left(|\widetilde{T}(\theta)\tilde{\varphi}|_{s,p} + |\tilde{\varphi}|_{t,p} \right). \tag{6.6}$$

The desired estimate (6.3) follows from estimate (6.6) with $s = 2 - 1/p$ and $t = -1/p$, just as in the proof of Theorem 4.12. □

(2) Now let u be an arbitrary function in the domain $D(\mathfrak{A})$:

$$u \in H^{2,p}(\Omega) \quad \text{and} \quad Bu = 0 \text{ on } \Gamma.$$

We choose a function $\zeta \in C^\infty(S)$ such that

$$\begin{cases} 0 \leq \zeta \leq 1 \quad \text{on } S, \\ \operatorname{supp} \zeta \subset \left[\frac{\pi}{3}, \frac{5\pi}{3}\right], \\ \zeta(y) = 1 \quad \text{for} \quad \frac{\pi}{2} \leq y \leq \frac{3\pi}{2}, \end{cases}$$

and let

$$\tilde{v}_\eta(x, y) = u(x) \otimes \zeta(y) e^{i\eta y}, \qquad \eta \geq 0.$$

Then we have

$$\begin{aligned} &\tilde{v}_\eta \in H^{2,p}(\Omega \times S), \\ &\widetilde{A}(\theta)\tilde{v}_\eta = \left(A + e^{i\theta} \frac{\partial^2}{\partial y^2}\right) \tilde{v}_\eta \\ &\quad = (A - \eta^2 e^{i\theta}) u \otimes \zeta e^{i\eta y} + 2(i\eta) e^{i\theta} u \otimes \zeta' e^{i\eta y} + e^{i\theta} u \otimes \zeta'' e^{i\eta y}, \end{aligned}$$

and also

$$B\tilde{v}_\eta = Bu \otimes \zeta e^{i\eta y} = 0 \quad \text{on } \Gamma \times S.$$

Thus, applying inequality (6.3) to the functions $\tilde{v}_\eta = u \otimes \zeta e^{i\eta y}$, we obtain that

$$(6.7) \qquad \left\| u \otimes \zeta e^{i\eta y} \right\|_{2,p} \leq \widetilde{C}(\theta) \left(\left\| \widetilde{A}(\theta)(u \otimes \zeta e^{i\eta y}) \right\|_p + \left\| u \otimes \zeta e^{i\eta y} \right\|_p \right).$$

We can estimate each term of inequality (6.7) as follows:

(6.8)

$$\left\| u \otimes \zeta e^{i\eta y} \right\|_p = \left(\int_{\Omega \times S} |u(x)|^p |\zeta(y)|^p \, dx \, dy \right)^{1/p} = \|\zeta\|_p \cdot \|u\|_p.$$

(6.9)

$$\begin{aligned} \left\| \widetilde{A}(\theta)(u \otimes \zeta e^{i\eta y}) \right\|_p &\leq \left\| (A - \eta^2 e^{i\theta}) u \otimes \zeta e^{i\eta y} \right\|_p \\ &\quad + 2\eta \left\| u \otimes \zeta' e^{i\eta y} \right\|_p + \left\| u \otimes \zeta'' e^{i\eta y} \right\|_p \\ &\leq \|\zeta\|_p \cdot \left\| (A - \eta^2 e^{i\theta}) u \right\|_p + (2\eta \|\zeta'\|_p + \|\zeta''\|_p) \|u\|_p. \end{aligned}$$

(6.10)

$$\begin{aligned} \left\| u \otimes \zeta e^{i\eta y} \right\|_{2,p}^p &= \sum_{|\alpha| \leq 2} \int_{\Omega \times S} \left| D_{x,y}^\alpha (u(x) \otimes \zeta(y) e^{i\eta y}) \right|^p \, dx dy \\ &\geq \sum_{|\alpha| \leq 2} \int_\Omega \int_{\pi/2}^{3\pi/2} \left| D_{x,y}^\alpha (u(x) \otimes e^{i\eta y}) \right|^p \, dx dy \\ &= \sum_{k + |\beta| \leq 2} \int_\Omega \int_{\pi/2}^{3\pi/2} \left| \eta^k D^\beta u(x) \right|^p \, dx dy \end{aligned}$$

$$\geq \pi\bigg(\sum_{|\beta|=2}\int_\Omega \left|D^\beta u(x)\right|^p\,dx + \eta^p \sum_{|\beta|=1}\int_\Omega \left|D^\beta u(x)\right|^p\,dx$$
$$+\,\eta^{2p}\int_\Omega |u(x)|^p\,dx\bigg)$$
$$= \pi\left(|u|_{2,p}^p + \eta^p|u|_{1,p}^p + \eta^{2p}\|u\|_p^p\right).$$

Therefore, carrying these inequalities (6.8), (6.9) and (6.10) into inequality (6.7), we have with a constant $\widetilde{C}'(\theta) > 0$ independent of $\eta > 0$:

$$|u|_{2,p} + \eta\,|u|_{1,p} + \eta^2\|u\|_p \leq \widetilde{C}'(\theta)\left(\left\|(A - \eta^2 e^{i\theta})u\right\|_p + \eta\,\|u\|_p\right).$$

If η is so large that

$$\eta \geq 2\widetilde{C}'(\theta),$$

then we can eliminate the last term on the right-hand side to obtain that

$$|u|_{2,p} + \eta\,|u|_{1,p} + \eta^2\|u\|_p \leq 2\widetilde{C}'(\theta)\left\|(A - \eta^2 e^{i\theta})u\right\|_p.$$

This proves inequality (6.2) if we take

$$\lambda = \eta^2 e^{i\theta},$$
$$R(\theta) = 4\widetilde{C}'(\theta)^2,$$
$$C(\theta) = 2\widetilde{C}'(\theta). \quad \square$$

The proof of Theorem 6.3 is now complete. $\square$

6.2 Generation of Analytic Semigroups

In this section we prove that the operator $\mathfrak{A}$ generates an analytic semigroup on the space $L^p(\Omega)$.

First we prove part (i) of Theorem 2:

Theorem 6.6. *Assume that conditions (H.1) and (H.2) are satisfied:*

(H.1) $a(x') \geq 0$ and $b(x') \geq 0$ on Γ.

(H.2) $b(x') > 0$ on $\Gamma_0 = \{x' \in \Gamma : a(x') = 0\}$.

Then, for every $0 < \varepsilon < \pi/2$, there exists a constant $r(\varepsilon) > 0$ such that the resolvent set of $\mathfrak{A}$ contains the set $\Sigma(\varepsilon) = \{\lambda = r^2 e^{i\theta} : r \geq r(\varepsilon), -\pi + \varepsilon \leq \theta \leq \pi - \varepsilon\}$, and that the resolvent $(\mathfrak{A} - \lambda I)^{-1}$ satisfies the estimate

$$\left\|(\mathfrak{A} - \lambda I)^{-1}\right\| \leq \frac{c(\varepsilon)}{|\lambda|}, \quad \lambda \in \Sigma(\varepsilon), \tag{6.11}$$

where $c(\varepsilon) > 0$ is a constant depending on ε.

Proof. (1) By estimate (6.2), it follows that if $\lambda = r^2 e^{i\theta}$, $-\pi < \theta < \pi$ and $|\lambda| = r^2 \geq R(\theta)$, then we have for all $u \in D(\mathfrak{A})$

$$|u|_{2,p} + |\lambda|^{1/2} \cdot |u|_{1,p} + |\lambda| \cdot \|u\|_p \leq C(\theta)\|(\mathfrak{A} - \lambda I)u\|_p.$$

But we find from the proof of Theorem 6.3 that the constants $R(\theta)$ and $C(\theta)$ depend *continuously* on $\theta \in (-\pi, \pi)$, so that they may be chosen uniformly in $\theta \in [-\pi + \varepsilon, \pi + \varepsilon]$, for every $\varepsilon > 0$. This proves the existence of the constants $r(\varepsilon)$ and $c(\varepsilon)$, that is, we have for all $\lambda = r^2 e^{i\theta}$ satisfying $r \geq r(\varepsilon)$ and $-\pi + \varepsilon \leq \theta \leq \pi + \varepsilon$

$$|u|_{2,p} + |\lambda|^{1/2} \cdot |u|_{1,p} + |\lambda| \cdot \|u\|_p \leq c(\varepsilon)\|(\mathfrak{A} - \lambda I)u\|_p. \tag{6.12}$$

By estimate (6.12), it follows that the operator $\mathfrak{A} - \lambda I$ is injective and its range $R(\mathfrak{A} - \lambda I)$ is closed in $L^p(\Omega)$, for all $\lambda \in \Sigma(\varepsilon)$.

(2) We show that the operator $\mathfrak{A} - \lambda I$ is surjective for all $\lambda \in \Sigma(\varepsilon)$:

$$R(\mathfrak{A} - \lambda I) = L^p(\Omega), \quad \lambda \in \Sigma(\varepsilon). \tag{6.13}$$

To do so, it suffices to show that the operator $\mathfrak{A} - \lambda I$ is a Fredholm operator with

$$\operatorname{ind}(\mathfrak{A} - \lambda I) = 0, \quad \lambda \in \Sigma(\varepsilon), \tag{6.14}$$

since $\mathfrak{A} - \lambda I$ is injective for all $\lambda \in \Sigma(\varepsilon)$.

(2-1) We reduce the study of the operator $\mathfrak{A} - \lambda I$ ($\lambda \in \Sigma(\varepsilon)$) to that of a pseudo-differential operator on the boundary, just as in the proof of Theorem 1.

Let $T(\lambda)$ be a classical pseudo-differential operator of first order on the boundary Γ defined as follows:

$$T(\lambda) = BP(\lambda) = a\Pi(\lambda) + b, \quad \lambda \in \Sigma(\varepsilon),$$

where

$$\begin{aligned} \Pi(\lambda) : C^\infty(\Gamma) &\longrightarrow C^\infty(\Gamma) \\ \varphi &\longmapsto \frac{\partial}{\partial \boldsymbol{\nu}}(P(\lambda)\varphi)\bigg|_\Gamma . \end{aligned}$$

Since the operator $T(\lambda) : C^\infty(\Gamma) \longrightarrow C^\infty(\Gamma)$ extends to a continuous linear operator $T(\lambda) : B^{\sigma,p}(\Gamma) \longrightarrow B^{\sigma-1,p}(\Gamma)$ for all $\sigma \in \mathbf{R}$, one can introduce a densely defined, closed linear operator

$$\mathcal{T}(\lambda) : B^{s-1/p,p}(\Gamma) \longrightarrow B^{s-1/p,p}(\Gamma)$$

as follows.

(α) The domain $D(\mathcal{T}(\lambda))$ of $\mathcal{T}(\lambda)$ is the space

$$D(\mathcal{T}(\lambda)) = \left\{\varphi \in B^{s-1/p,p}(\Gamma) : T(\lambda)\varphi \in B^{s-1/p,p}(\Gamma)\right\}.$$

(β) $\mathcal{T}(\lambda)\varphi = T(\lambda)\varphi$, $\varphi \in D(\mathcal{T}(\lambda))$.

Then we can obtain the following results (cf. Theorem 4.10):

(I) The null space $N(\mathfrak{A} - \lambda I)$ of $\mathfrak{A} - \lambda I$ has finite dimension if and only if the null space $N(\mathcal{T}(\lambda))$ of $\mathcal{T}(\lambda)$ has finite dimension, and we have

$$\dim N(\mathfrak{A} - \lambda I) = \dim N(\mathcal{T}(\lambda)).$$

(II) The range $R(\mathfrak{A} - \lambda I)$ of $\mathfrak{A} - \lambda I$ is closed if and only if the range $R(\mathcal{T}(\lambda))$ of $\mathcal{T}(\lambda)$ is closed; and $R(\mathfrak{A} - \lambda I)$ has finite codimension if and only if $R(\mathcal{T}(\lambda))$ has finite codimension, and we have

$$\operatorname{codim} R(\mathfrak{A} - \lambda I) = \operatorname{codim} R(\mathcal{T}(\lambda)).$$

(III) The operator $\mathfrak{A}-\lambda I$ is a Fredholm operator if and only if the operator $\mathcal{T}(\lambda)$ is a Fredholm operator, and we have

$$\operatorname{ind}(\mathfrak{A} - \lambda I) = \operatorname{ind} \mathcal{T}(\lambda).$$

Therefore, assertion (6.4) is reduced to the following assertion:

$$\operatorname{ind} \mathcal{T}(\lambda) = 0, \quad \lambda \in \Sigma(\varepsilon). \tag{6.14$'$}$$

(2-2) To prove assertion (6.14$'$), we shall make use of the method of Agmon just as in Section 5.3.

Let $\widetilde{T}(\theta)$ be the classical pseudo-differential operator of first order on the boundary $\Gamma \times S$ introduced in Section 6.1:

$$\widetilde{T}(\theta) = B\widetilde{P}(\theta) = a\widetilde{\Pi}(\theta) + b, \quad -\pi < \theta < \theta,$$

where

$$\widetilde{\Pi}(\theta) : C^\infty(\Gamma \times S) \longrightarrow C^\infty(\Gamma \times S)$$

$$\tilde{\varphi} \longmapsto \frac{\partial}{\partial \boldsymbol{\nu}}\left(\widetilde{P}(\theta)\tilde{\varphi}\right)\Big|_{\Gamma \times S}.$$

We define a densely defined, closed linear operator

$$\widetilde{\mathcal{T}}(\theta) : B^{s-1/p,p}(\Gamma \times S) \longrightarrow B^{s-1/p,p}(\Gamma \times S)$$

as follows.

($\tilde{\alpha}$) The domain $D\left(\widetilde{\mathcal{T}}(\theta)\right)$ of $\widetilde{\mathcal{T}}(\theta)$ is the space

$$D\left(\widetilde{\mathcal{T}}(\theta)\right) = \left\{\tilde{\varphi} \in B^{s-1/p,p}(\Gamma \times S) : \widetilde{T}(\theta)\tilde{\varphi} \in B^{s-1/p,p}(\Gamma \times S)\right\}.$$

($\tilde{\beta}$) $\widetilde{\mathcal{T}}(\theta)\tilde{\varphi} = \widetilde{T}(\theta)\tilde{\varphi}$, $\tilde{\varphi} \in D\left(\widetilde{\mathcal{T}}(\theta)\right)$.

Then we can prove the the most fundamental relationship between the operators $\widetilde{\mathcal{T}}(\theta)$ and $\mathcal{T}(\lambda)$, analogous to Proposition 5.10:

Proposition 6.7. *If* $\operatorname{ind}\widetilde{T}(\theta)$ *is finite, then there exists a finite subset* K *of* $\mathbf{Z}$ *such that the operator* $T(\lambda')$ *is bijective for all* $\lambda' = \ell^2 e^{i\theta}$ *satisfying* $\ell \in \mathbf{Z} \setminus K$.

(3) *End of proof of Theorem 6.6*

(3-1) We show that if conditions (H.1) and (H.2) are satisfied, then we have

$$\operatorname{ind}\widetilde{T}(\theta) < \infty. \tag{6.15}$$

Now, estimate (6.6) with $s = s - 1/p$ gives that

$$|\tilde{\varphi}|_{s-1/p,p} \leq \widetilde{C}_{s,t}\left(|\widetilde{T}(\theta)\tilde{\varphi}|_{s-1/p,p} + |\tilde{\varphi}|_{t,p}\right), \quad \tilde{\varphi} \in D(\widetilde{T}(\theta)), \tag{6.16}$$

where $t < s - 1/p$. But it follows from an application of Rellich's theorem that the injection $B^{s-1/p,p}(\Gamma \times S) \longrightarrow B^{t,p}(\Gamma \times S)$ is *compact* for $t < s - 1/p$. Thus, applying Lemma 5.11 with

$$\begin{aligned} X = Y &= B^{s-1/p,p}(\Gamma \times S), \\ Z &= B^{t,p}(\Gamma \times S), \\ T &= \widetilde{T}(\theta), \end{aligned}$$

we obtain that the range $R\left(\widetilde{T}(\theta)\right)$ is closed in $B^{s-1/p,p}(\Gamma \times S)$ and

$$\dim N\left(\widetilde{T}(\theta)\right) < \infty.$$

On the other hand, by formula (6.5), we find that the complete symbol of the adjoint $\widetilde{T}(\theta)^*$ is given by the following (cf. Theorem 3.12):

$$\begin{aligned} & a(x')\left(\tilde{p}_1(x',\xi',y,\eta;\theta) - \sqrt{-1}\,\tilde{q}_1(x',\xi',y,\eta;\theta)\right) \\ & + \left(\left[b(x') + a(x')\tilde{p}_0(x',\xi',y,\eta;\theta) - \sum_{j=1}^{n-1} \partial_{x_j}\left(a(x')\cdot \partial_{\xi_j}\tilde{q}_1(x',\xi',y,\eta;\theta)\right)\right]\right. \\ & \qquad \left. - \sqrt{-1}\left[a(x')\tilde{q}_0(x',\xi',y,\eta;\theta) + \sum_{j=1}^{n-1} \partial_{x_j}\left(a(x')\cdot \partial_{\xi_j}\tilde{p}_1(x',\xi',y,\eta;\theta)\right)\right]\right) \\ & + \text{terms of order} \leq -1. \end{aligned}$$

But, by virtue of Lemma 5.4, it follows that

$$\partial_{x_j} a(x') = 0 \text{ on } \Gamma_0 = \{x' \in \Gamma : a(x') = 0\}.$$

Thus one can easily verify that the pseudo-differential operator $\widetilde{T}(\theta)^*$ satisfies conditions (3.7a) and (3.7b) of Theorem 3.19 with $\mu = 0$, $\rho = 1$ and $\delta = 1/2$.

This implies that estimate (6.16) remains valid for the adjoint operator $\widetilde{\mathcal{T}}(\theta)^*$ of $\widetilde{\mathcal{T}}(\theta)$:

$$|\tilde{\psi}|_{-s+1/p,p'} \leq \widetilde{C}_{s,\tau}\left(|\widetilde{\mathcal{T}}(\theta)^*\tilde{\psi}|_{-s+1/p,p'} + |\tilde{\psi}|_{\tau,p'}\right), \quad \tilde{\psi} \in D\left(\widetilde{\mathcal{T}}(\theta)^*\right),$$

where $\tau < -s + 1/p$ and $p' = p/(p-1)$, the exponent conjugate to p.

Therefore, assertion (6.5) can be proved just as in the proof of Proposition 5.8.

(3-2) By assertion (6.15), we can apply Proposition 6.7 to obtain that the operator $\mathcal{T}(\ell^2 e^{i\theta}) : B^{s-1/p,p}(\Gamma) \longrightarrow B^{s-1/p,p}(\Gamma)$ is bijective if $\ell \in \mathbf{Z} \setminus K$ for some finite subset K of $\mathbf{Z}$. In particular, we have

$$\operatorname{ind} \mathcal{T}(\ell^2 e^{i\theta}) = 0 \quad \text{if } \ell \in \mathbf{Z} \setminus K.$$

Thus, just as in the proof of Proposition 5.8, we can prove assertion (6.14′) and hence assertion (6.13).

(3-3) Summing up, we have proved that the operator $\mathfrak{A} - \lambda I$ is bijective for all $\lambda \in \Sigma(\varepsilon)$ and its inverse $(\mathfrak{A} - \lambda I)^{-1}$ satisfies estimate (6.11).

The proof of Theorem 6.6 (part (i) of Theorem 2) is complete. □

Part (ii) of Theorem 2 may be proved as follows. Theorem 6.6 tells us that, for $\mu_\varepsilon > 0$ large enough, the operator $\mathfrak{A} - \mu_\varepsilon I$ satisfies condition (1.1) (see Figure 6.1).

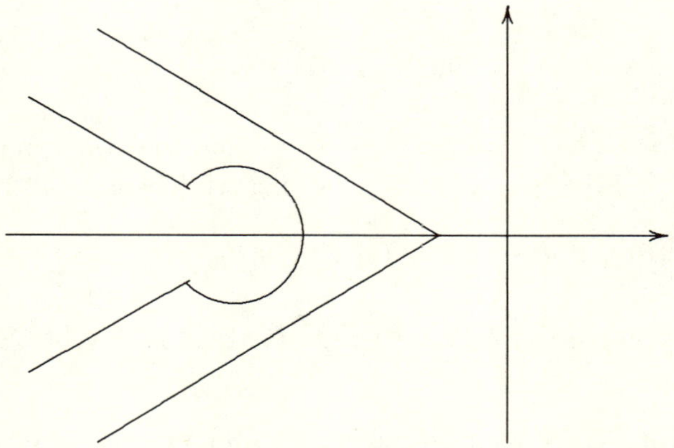

Figure 6.1

Thus, applying Theorem 1.2 (and Remark 1.3) to the operator $\mathfrak{A} - \mu_\varepsilon I$, we obtain part (ii) of Theorem 2:

Theorem 6.8. *If conditions (H.1) and (H.2) are satisfied, then the operator $\mathfrak{A}$ generates a semigroup $U(z)$ on $L^p(\Omega)$ which is analytic in the sector*

$$\Delta_\varepsilon = \{z = t + is : z \neq 0, |\arg z| < \pi/2 - \varepsilon\}$$

for any $0 < \varepsilon < \pi/2$, and enjoys the following properties:

(a) The operators $\mathfrak{A}U(z)$ and $\frac{dU}{dz}(z)$ are bounded operators on $L^p(\Omega)$ for each $z \in \Delta_\varepsilon$, and satisfy the relation

$$\frac{dU}{dz}(z) = \mathfrak{A}U(z), \quad z \in \Delta_\varepsilon.$$

(b) For each $0 < \varepsilon < \pi/2$, there exist constants $\widetilde{M_0}(\varepsilon) > 0$, $\widetilde{M_1}(\varepsilon) > 0$ and $\mu_\varepsilon > 0$ such that

$$\|U(z)\| \leq \widetilde{M_0}(\varepsilon) e^{\mu_\varepsilon \cdot \mathrm{Re}\, z}, \quad z \in \Delta_\varepsilon,$$

$$\|\mathfrak{A}U(z)\| \leq \frac{\widetilde{M_1}(\varepsilon)}{|z|} e^{\mu_\varepsilon \cdot \mathrm{Re}\, z}, \quad z \in \Delta_\varepsilon.$$

(c) For each $u_0 \in L^p(\Omega)$, we have as $z \to 0, z \in \Delta_\varepsilon$,

$$U(z)u_0 \longrightarrow u_0 \quad \text{in } L^p(\Omega).$$

The proof of Theorem 2 is now complete. □

CHAPTER VII

PROOF OF THEOREMS 3 AND 4

This chapter is devoted to the semigroup approach to the following *semi-linear* initial boundary value problem: Given functions f and u_0 defined in $\Omega \times [0,T) \times \mathbf{R} \times \mathbf{R}^N$ and in Ω, respectively, find a function u in $\Omega \times [0,T)$ such that

(**)
$$\begin{cases} (\frac{\partial}{\partial t} - A)u(x,t) = f(x,t,u,\operatorname{grad} u) & \text{in } \Omega \times (0,T), \\ Bu(x',t) = a(x')\frac{\partial u}{\partial \nu}(x',t) + b(x')u(x',t)\big|_{\Gamma\times[0,T)} = 0 & \text{on } \Gamma \times [0,T), \\ u(x,0) = u_0(x) & \text{in } \Omega. \end{cases}$$

By using the operator $\mathfrak{A}$, one can formulate problem (**) in terms of the *abstract Cauchy problem* in the Banach space $L^p(\Omega)$ as follows:

$$(**') \qquad \begin{cases} \frac{du}{dt} = \mathfrak{A}u(t) + F(t,u(t)), \ 0 < t < T, \\ u|_{t=0} = u_0. \end{cases}$$

Here $u(t) = u(\cdot,t)$ and $F(t,u(t)) = f(\cdot,t,u(t),\operatorname{grad} u(t))$ are functions defined on the interval $[0,T)$, taking values in the space $L^p(\Omega)$.

Our semigroup approach can be traced back to the pioneering work of Fujita and Kato [FK]. For detailed studies of semilinear parabolic equations, the reader is referred to Friedman [Fr] and Henry [He].

7.1 Fractional Powers and Imbedding Theorems

By Theorem 6.6, one may assume that the operator $\mathfrak{A}$ satisfies condition (1.17).

(1) The resolvent set of $\mathfrak{A}$ contains the region Σ shown in Figure 7.1.

(2) There exists a constant $M > 0$ such that the resolvent $R(\lambda) = (\mathfrak{A} - \lambda I)^{-1}$ satisfies the estimate

$$(7.1) \qquad \|R(\lambda)\| \leq \frac{M}{1+|\lambda|}, \quad \lambda \in \Sigma.$$

Thus we can define the fractional powers $(-\mathfrak{A})^\alpha$ for $0 < \alpha < 1$ on the space $L^p(\Omega)$ as follows (cf. formula (1.19)):

$$(-\mathfrak{A})^{-\alpha} = -\frac{\sin \alpha\pi}{\pi} \int_0^\infty s^{-\alpha}(\mathfrak{A} - sI)^{-1}\, ds,$$

Typeset by $\mathcal{A}_{\mathcal{M}}\mathcal{S}$-TEX

and

$$(-\mathfrak{A})^{\alpha} = \text{the inverse of } (-\mathfrak{A})^{-\alpha}, \quad \alpha > 0.$$

Recall that the operator $(-\mathfrak{A})^{\alpha}$ is a closed linear operator with domain $D((-\mathfrak{A})^{\alpha}) \supset D(\mathfrak{A})$.

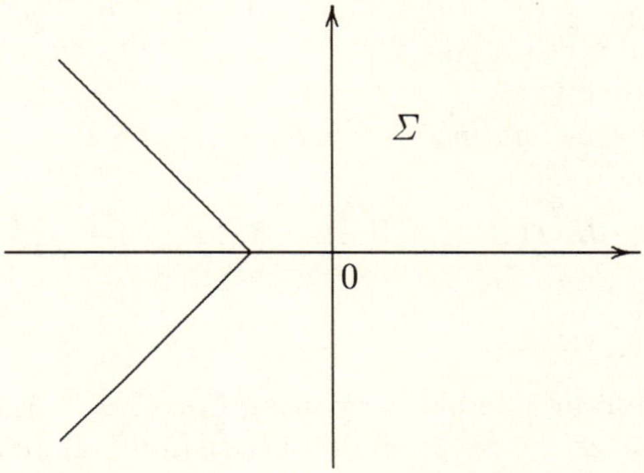

Figure 7.1

In this section we study the imbedding characteristics of the spaces $D((-\mathfrak{A})^{\alpha})$, which will make these spaces so useful in the study of semilinear parabolic differential equations.

We let

X_{α} = the space $D((-\mathfrak{A})^{\alpha})$ endowed with the graph norm $\|\cdot\|_{\alpha}$ of $(-\mathfrak{A})^{\alpha}$.

Here

$$\|u\|_{\alpha} = \left(\|u\|_p^2 + \|(-\mathfrak{A})^{\alpha} u\|_p^2\right)^{1/2}, \quad u \in D((-\mathfrak{A})^{\alpha}).$$

Then we have the following (cf. Proposition 1.17):

(1) The space X_{α} is a Banach space.
(2) The graph norm $\|u\|_{\alpha}$ is equivalent to the norm $\|(-\mathfrak{A})^{\alpha} u\|_p$.
(3) If $0 < \alpha < \beta < 1$, then we have $X_{\beta} \subset X_{\alpha}$ with continuous injection.

The next theorem gives the imbedding properties of the spaces X_{α} into the Sobolev spaces (cf. [He, Theorem 1.6.1]):

Theorem 7.1. *Let $1 < p < \infty$. Then we have the following continuous injections:*

(i) $X_{\alpha} \subset H^{1,q}(\Omega)$ *if* $\frac{1}{2} < \alpha < 1$, $\frac{1}{p} - \frac{2\alpha - 1}{n} < \frac{1}{q} \leq \frac{1}{p}$, $1 < p < n$.
(ii) $X_{\alpha} \subset C^{\nu}(\overline{\Omega})$ *if* $\frac{n}{2p} < \alpha < 1$, $0 \leq \nu < 2\alpha - \frac{n}{p}$, $p \neq n$.

The proof of Theorem 7.1 is based on the following two results which characterize the imbedding properties of Sobolev spaces:

Theorem 7.2 (Sobolev). *Let Ω be a bounded domain in $\mathbf{R}^n$ with boundary Γ of class C^2. Then:*

(i) If $1 \le p < n$, we have

$$H^{2,p}(\Omega) \subset H^{1,q}(\Omega), \quad \frac{1}{p} - \frac{1}{n} \le \frac{1}{q} \le \frac{1}{p},$$

with continuous injection.

(ii) If $n/2 < p < \infty$ and $p \ne n$, we have

$$H^{2,p}(\Omega) \subset C^{\nu}(\overline{\Omega}), \quad 0 < \nu \le 2 - \frac{n}{p},$$

with continuous injection.

Part (i) of Theorem 7.2 follows by using Corollary 2.22 and Theorem 2.25 with $m = 2$, $j = 1$ and $r := q$, while part (ii) of Theorem 7.2 follows from Theorem 2.18 and Theorem 2.25 with $m = 2$ (cf. [Ad, Theorem 5.4]).

Theorem 7.3 (Gagliardo-Nirenberg). *Let Ω be a bounded domain in $\mathbf{R}^n$ with boundary of class C^2, and $1 \le p, r \le \infty$. Then:*

(i) If $p \ne n$ and if

$$\frac{1}{q} = \frac{1}{n} + \theta\left(\frac{1}{p} - \frac{2}{n}\right) + (1-\theta)\frac{1}{r}, \quad \frac{1}{2} \le \theta \le 1,$$

then we have for all $u \in H^{2,p}(\Omega) \cap L^r(\Omega)$

$$\|u\|_{1,q} \le C_1 \|u\|_{2,p}^{\theta} \|u\|_r^{1-\theta}, \tag{7.2}$$

with a constant $C_1 = C_1(\Omega, p, r, \theta) > 0$.

(ii) If $n/2 < p < \infty$, $p \ne n$ and if

$$0 \le \nu < \theta\left(2 - \frac{n}{p}\right) - (1-\theta)\frac{n}{r},$$

then we have for all $u \in H^{2,p}(\Omega) \cap L^r(\Omega)$

$$\|u\|_{C^{\nu}(\overline{\Omega})} \le C_2 \|u\|_{2,p}^{\theta} \|u\|_r^{1-\theta}, \tag{7.3}$$

with a constant $C_2 = C_2(\Omega, p, r, \theta) > 0$.

Part (i) of Theorem 7.3 follows by using from Theorem 2.15 and Theorem 2.25 with $m = 2$, $j = 1$, $q := r$ and $r := q$, while part (ii) of Theorem 7.3 follows from Theorem 2.18 and Theorem 2.25 with $m = 2$, $j = \nu$, $q := r$ and $r := \infty$ (cf. [Fr, Part I, Theorem 10.1]).

Proof of Theorem 7.1. First, by estimate (7.1) with $\lambda = 0$, it follows that

$$\|u\|_{2,p} \leq M\|\mathfrak{A}u\|_p, \quad u \in D(\mathfrak{A}). \tag{7.1$'$}$$

(i) For each $\theta \in (1/2, 1)$, we let

$$\frac{1}{r} = \frac{1}{p} + \frac{\varepsilon}{1-\theta}, \quad 0 < \varepsilon \leq \frac{2\theta - 1}{n},$$

and

$$\frac{1}{q} = \frac{1}{p} - \frac{2\theta - 1}{n} + \varepsilon = \frac{1}{n} + \theta\left(\frac{1}{p} - \frac{2}{n}\right) + (1-\theta)\frac{1}{r}.$$

Then it follows from inequalities (7.2) and (7.1$'$) that

$$\begin{aligned} \|u\|_{1,q} &\leq C_1 \|u\|_{2,p}^{\theta} \|u\|_r^{1-\theta} \\ &\leq MC_1 \|\mathfrak{A}u\|_p^{\theta} \|u\|_p^{1-\theta}. \end{aligned}$$

Hence we have for all $\delta > 0$

$$\|u\|_{1,q} \leq MC_1\left(\delta^{-\theta}\|u\|_p + \delta^{1-\delta}\|\mathfrak{A}u\|_p\right), \quad u \in D(\mathfrak{A}). \tag{7.4}$$

But, if we let

$$B_1 = \text{the identity operator on the space } H^{1,q}(\Omega),$$

then, by part (i) of Theorem 7.2, it follows that

$$D(\mathfrak{A}) \subset H^{2,p}(\Omega) \subset H^{1,q}(\Omega) = D(B_1),$$

since we have

$$\frac{1}{p} - \frac{2\theta - 1}{n} < \frac{1}{q} \leq \frac{1}{p}, \quad \frac{1}{2} < \theta < 1.$$

Further we can write inequality (7.4) as follows:

$$\|B_1 u\|_{1,q} \leq MC_1\left(\delta^{-\theta}\|u\|_p + \delta^{1-\theta}\|\mathfrak{A}u\|_p\right), \quad u \in D(\mathfrak{A}). \tag{7.4$'$}$$

Therefore, applying Lemma 1.13 to our situation, we obtain that

$$X_\alpha = D((-\mathfrak{A})^\alpha) \subset D(B_1) = H^{1,q}(\Omega), \quad \theta < \alpha < 1,$$

and

$$\|u\|_{1,q} \leq K_\alpha \|(-\mathfrak{A})^\alpha u\|_p, \quad u \in X_\alpha.$$

This proves part (i).

(ii) For each $\theta \in (n/2p, 1)$, we let

$$r = p,$$

and

$$0 < \nu < 2\theta - \frac{n}{p} = \theta\left(2 - \frac{n}{p}\right) - (1-\theta)\frac{n}{r}.$$

Then it follows from inequalities (7.3) and (7.1′) that

$$\begin{aligned}\|u\|_{C^\nu(\overline{\Omega})} &\leq C_2 \|u\|_{2,p}^{\theta} \|u\|_p^{1-\theta} \\ &\leq M C_2 \|\mathfrak{A} u\|_p^{\theta} \|u\|_p^{1-\theta}.\end{aligned}$$

Hence we have for all $\delta > 0$

$$\text{(7.5)} \qquad \|u\|_{C^\nu(\overline{\Omega})} \leq M C_2 \left(\delta^{-\theta}\|u\|_p + \delta^{1-\delta}\|\mathfrak{A}u\|_p\right), \quad u \in D(\mathfrak{A}).$$

But, if we let

$$B_2 = \text{the identity operator on the space } C^\nu(\overline{\Omega}),$$

then, by part (ii) of Theorem 7.2, it follows that

$$D(\mathfrak{A}) \subset H^{2,p}(\Omega) \subset C^\nu(\overline{\Omega}) = D(B_2),$$

since we have

$$0 < \nu < 2\theta - \frac{n}{p}, \quad \frac{1}{2} < \theta < 1.$$

Further we can write inequality (7.5) as follows:

$$\text{(7.5′)} \qquad \|B_2 u\|_{C^\nu(\overline{\Omega})} \leq M C_2 \left(\delta^{-\theta}\|u\|_p + \delta^{1-\theta}\|\mathfrak{A}u\|_p\right), \quad u \in D(\mathfrak{A}).$$

Therefore, applying Lemma 1.13 to our situation, we obtain that

$$X_\alpha = D((-\mathfrak{A})^\alpha) \subset D(B_2) = C^\nu(\overline{\Omega}), \quad \theta < \alpha < 1,$$

and

$$\|u\|_{C^\nu(\overline{\Omega})} \leq K'_\alpha \left\|(-\mathfrak{A})^\alpha u\right\|_p, \quad u \in X_\alpha.$$

This proves part (ii).

The proof of Theorem 7.1 is complete. □

7.2 Semilinear Initial Boundary Value Problems

This final section is devoted to the proof of Theorems 3 and 4.

7.2A Proof of Theorem 3.

We verify that all the conditions of Theorem 1.18 are satisfied; then Theorem 3 follows from an application of the same theorem.

Since $p > n$, one can choose a constant α such that

$$\frac{1}{2}\left(\frac{n}{p}+1\right) < \alpha < 1,$$

so that

$$1 < 2\alpha - \frac{n}{p}.$$

Then, by part (ii) of Theorem 7.1 with $\nu = 1$, we have

$$X_\alpha \subset C^1(\overline{\Omega}) \quad \text{and} \quad X_\alpha \subset H^{1,p}(\Omega), \tag{7.6}$$

with continuous injections. Thus we find that the function

$$F(t,u) := f(x,t,u(x),\operatorname{grad} u(x))$$

is well defined on $[0,T] \times X_\alpha$. Furthermore, since the function $f(x,t,u,\xi)$ is locally Lipschitz continuous, in view of assertion (7.6) it follows that, for all $t, s \in [0, t_0]$ and for all $u, v \in X_\alpha$ with $\|u - u_0\|_\alpha \leq R$, $\|v - u_0\|_\alpha \leq R$,

$$\begin{aligned} \|F(t,u) - F(s,v)\|_p &\leq \|F(t,u) - F(t,v)\|_p + \|F(t,v) - F(s,v)\|_p \\ &\leq C\left(\|u-v\|_p + \sum_{i=1}^n \left\|\frac{\partial}{\partial x_i}(u-v)\right\|_p + |t-s|\right) \\ &\leq C\left(\|u-v\|_{1,p} + |t-s|\right) \\ &\leq CC'\left(\|u-v\|_\alpha + |t-s|\right). \end{aligned} \tag{7.7}$$

Here $C = C(t_0, R) > 0$ is a (local) Lipschitz constant for the function f, and $C' > 0$ is an imbedding constant for the imbedding $X_\alpha \subset H^{1,p}(\Omega)$.

By inequality (7.7), we obtain that the function $F(t,u)$ is locally Lipschitz continuous in t and u.

The proof of Theorem 3 is complete. □

7.2B Proof of Theorem 4.

The proof is similar to that of Theorem 3; we verify that all the conditions of Theorem 1.18 are satisfied.

Since $n/2 < p < n$ and $1 \leq \gamma < n/(n-p)$, one can choose a constant α such that

$$\max\left(\frac{n}{2p}, \frac{1}{2} + \frac{n}{2p}\left(\frac{\gamma-1}{\gamma}\right)\right) < \alpha < 1, \tag{7.8}$$

so that

$$0 < 2\alpha - \frac{n}{p} \quad \text{and} \quad \frac{1}{p} - \frac{2\alpha - 1}{n} < \frac{1}{p\gamma} \leq \frac{1}{p}.$$

Then, by Theorem 7.1 with $\nu = 0$ and $q = p\gamma$, we have

$$X_\alpha \subset L^\infty(\Omega) \quad \text{and} \quad X_\alpha \subset H^{1,p\gamma}(\Omega), \tag{7.9}$$

with continuous injections.

We let

$$F(t,u) := f(x,t,u(x), \operatorname{grad} u(x)), \quad t \in [0,T], \ u \in X_\alpha.$$

Then we have by condition (a) of Theorem 4

$$\begin{aligned}\|F(t,u)\|_p^p &\leq 2^{p-1} \rho(t, \|u\|_\infty)^p \int_\Omega (1 + |\operatorname{grad} u|^{p\gamma})\, dx \\ &\leq 2^{p-1} \rho(t, \|u\|_\infty)^p \left(|\Omega| + \|u\|_{1,p\gamma}^{p\gamma}\right).\end{aligned}$$

Here and in the following, $|\Omega|$ denotes the volume of the domain Ω. By assertion (7.9), it follows that the function $F(t,u)$ is well defined on $[0,T] \times X_\alpha$ for all α satisfying condition (7.8).

(1) First we verify the local Lipschitz continuity of $F(t,u)$ with respect to the variable t.

By condition (b) of Theorem 4, it follows that

$$\begin{aligned}&\|F(t,u) - F(s,u)\|_p^p \\ &= \int_\Omega |f(x,t,u(x), \operatorname{grad} u(x)) - f(x,s,u(x), \operatorname{grad} u(x))|^p\, dx \\ &\leq 2^{p-1} \rho(t, \|u\|_\infty)^p |t-s|^p \int_\Omega (1 + |\operatorname{grad} u|^{p\gamma})\, dx \\ &\leq 2^{p-1} \rho(t, \|u\|_\infty)^p \left(|\Omega| + \|u\|_{1,p\gamma}^{p\gamma}\right) |t-s|^p.\end{aligned}$$

In view of assertion (7.9), this proves that

$$\|F(t,u) - F(s,u)\|_p \leq C_1\left(\|u\|_\alpha\right) |t-s|, \tag{7.10}$$

where $C_1(\|u\|_\alpha) > 0$ is a constant depending on the norm $\|u\|_\alpha$.

(2) Next we verify the local Lipschitz continuity of $F(t,u)$ with respect to the variable u.

To do so, we remark that

$$\begin{aligned}&\|F(t,u) - F(t,v)\|_p^p \\ &= \int_\Omega |f(x,t,u(x), \operatorname{grad} u(x)) - f(x,t,v(x), \operatorname{grad} v(x))|^p\, dx \\ &\leq 2^{p-1} \int_\Omega |f(x,t,u(x), \operatorname{grad} u(x)) - f(x,t,u(x), \operatorname{grad} v(x))|^p\, dx \\ &\quad + 2^{p-1} \int_\Omega |f(x,t,u(x), \operatorname{grad} v(x)) - f(x,t,v(x), \operatorname{grad} v(x))|^p\, dx.\end{aligned} \tag{7.11}$$

We estimate the two terms on the right of inequality (7.11).

(2-1) By condition (c) of Theorem 4, it follows that

(7.12)
$$\int_\Omega |f(x,t,u(x),\operatorname{grad} u(x)) - f(x,t,u(x),\operatorname{grad} v(x))|^p\,dx$$
$$\leq 3^{p-1}\rho(t,\|u\|_\infty)^p \int_\Omega (1 + |\operatorname{grad} u|^{p(\gamma-1)} + |\operatorname{grad} v|^{p(\gamma-1)})|\operatorname{grad}(u-v)|^p\,dx.$$

But, by Hölder's inequality, it follows that

(7.13)
$$\int_\Omega |\operatorname{grad}(u-v)|^p\,dx \leq \left(\int_\Omega 1\,dx\right)^{(\gamma-1)/\gamma} \left(\int_\Omega |\operatorname{grad}(u-v)|^{p\gamma}\,dx\right)^{1/\gamma}$$
$$\leq |\Omega|^{(\gamma-1)/\gamma}\|u-v\|_{1,p\gamma}^p,$$

and also

$$\int_\Omega |\operatorname{grad} u|^{p(\gamma-1)}|\operatorname{grad}(u-v)|^p\,dx \tag{7.14}$$
$$\leq \left(\int_\Omega |\operatorname{grad} u|^{p\gamma}\right)^{(\gamma-1)/\gamma} \left(\int_\Omega |\operatorname{grad}(u-v)|^{p\gamma}\,dx\right)^{1/\gamma}$$
$$\leq \|u\|_{1,p\gamma}^{p(\gamma-1)}\|u-v\|_{1,p\gamma}^p$$

and

$$\int_\Omega |\operatorname{grad} v|^{p(\gamma-1)}\,|\operatorname{grad}(u-v)|^p\,dx \leq \|v\|_{1,p\gamma}^{p(\gamma-1)}\,\|u-v\|_{1,p\gamma}^p. \tag{7.15}$$

Thus, carrying inequalities (7.13), (7.14) and (7.15) into inequality (7.12), we obtain that

(7.16)
$$\int_\Omega |f(x,t,u(x),\operatorname{grad} u(x)) - f(x,t,u(x),\operatorname{grad} v(x))|^p\,dx$$
$$\leq 3^{p-1}\rho(t,\|u\|_\infty)^p \left(|\Omega|^{(\gamma-1)/\gamma} + \|u\|_{1,p\gamma}^{p(\gamma-1)} + \|v\|_{1,p\gamma}^{p(\gamma-1)}\right)\|u-v\|_{1,p\gamma}^p.$$

(2-2) By condition (d) of Theorem 4, it follows that

$$\int_\Omega |f(x,t,u(x),\operatorname{grad} v(x)) - f(x,t,v(x),\operatorname{grad} v(x))|^p\,dx \tag{7.17}$$
$$\leq 2^{p-1}\rho(t,\|u\|_\infty + \|v\|_\infty)^p \int_\Omega (1 + |\operatorname{grad} v|^{p\gamma})\,|u-v|^p\,dx$$
$$\leq 2^{p-1}\rho(t,\|u\|_\infty + \|v\|_\infty)^p\|u-v\|_\infty^p \int_\Omega (1 + |\operatorname{grad} v|^{p\gamma})\,dx$$
$$\leq 2^{p-1}\rho(t,\|u\|_\infty + \|v\|_\infty)^p \left(|\Omega| + \|v\|_{1,p\gamma}^{p\gamma}\right)\|u-v\|_\infty^p.$$

Therefore, combining inequalities (7.11), (7.16) and (7.17), we obtain that

$$\begin{aligned} &\|F(t,u) - F(t,v)\|_p^p \\ &\leq 6^{p-1}\rho(t, \|u\|_\infty)^p \left(|\Omega|^{(\gamma-1)/\gamma} + \|u\|_{1,p\gamma}^{p(\gamma-1)} + \|v\|_{1,p\gamma}^{p(\gamma-1)}\right) \|u - v\|_{1,p\gamma}^p \\ &\quad + 4^{p-1}\rho(t, \|u\|_\infty + \|v\|_\infty)^p \left(|\Omega| + \|v\|_{1,p\gamma}^{p\gamma}\right) \|u - v\|_\infty^p. \end{aligned}$$

In view of assertion (7.9), this proves that

$$\|F(t,u) - F(t,v)\|_p \leq C_2(\|u\|_\alpha, \|v\|_\alpha)\, \|u - v\|_\alpha, \tag{7.18}$$

where $C_2(\|u\|_\alpha, \|v\|_\alpha) > 0$ is a constant depending on the norms $\|u\|_\alpha$ and $\|v\|_\alpha$.

Summing up, we find from inequalities (7.10) and (7.18) that the function $F(t,u)$ is locally Lipschitz continuous in t and u.

The proof of Theorem 4 is now complete. □

APPENDIX: THE MAXIMUM PRINCIPLE

Let Ω be a bounded domain of Euclidean space $\mathbf{R}^n$, with boundary Γ, and let A be an elliptic second-order differential operator with real coefficients such that

$$A = \sum_{i,j=1}^{n} a^{ij}(x) \frac{\partial^2}{\partial x_i \partial x_j} + \sum_{i=1}^{n} b^i(x) \frac{\partial}{\partial x_i} + c(x)$$

where:

(1) $a^{ij} \in C(\mathbf{R}^n)$, $a^{ij} = a^{ji}$, $1 \leq i, j \leq n$ and there exists a constant $a_0 > 0$ such that

$$\sum_{i,j=1}^{n} a^{ij}(x)\xi_i\xi_j \geq a_0|\xi|^2, \quad x \in \mathbf{R}^n, \ \xi = (\xi_1, \xi_2, \cdots, \xi_n) \in \mathbf{R}^n.$$

(2) $b^i \in C(\mathbf{R}^n)$.

(3) $c \in C(\mathbf{R}^n)$ and $c \leq 0$ in Ω.

First we have the following strong maximum principle:

Theorem A.1 (The strong maximum principle). *Assume that*

$$\begin{cases} u \in C^2(\overline{\Omega}), \ Au \geq 0 \quad \text{in } \Omega, \\ m = \max_{\overline{\Omega}} u \geq 0. \end{cases}$$

If the function u takes its maximum m at some point x_0 of the interior *Ω, then $u \equiv m$ in the connected component containing x_0.*

Now assume that Ω is a *domain of class C^2*, that is, each point of the boundary Γ has a neighborhood in which Γ is the graph of a C^2 function of $n-1$ of the variables $x_1, x_2, \cdots, x_n$. We consider a function $u \in C(\overline{\Omega}) \cap C^2(\Omega)$ which satisfies the condition

$$Au \geq 0 \quad \text{in } \Omega,$$

and study the exterior normal derivative $\partial u/\partial \boldsymbol{\nu}$ at a point where the function u takes its non-negative maximum.

The boundary point lemma reads as follows:

Typeset by $\mathcal{A}\mathcal{M}\mathcal{S}$-TeX

Theorem A.2 (The boundary point lemma). *Let Ω be a domain of class C^2. Assume that a function $u \in C(\overline{\Omega}) \cap C^2(\Omega)$ satisfies the condition*

$$Au \geq 0 \quad \text{in } \Omega,$$

and that there exists a point $x_0' \in \Gamma$ such that

$$\begin{cases} u(x_0') = \max_{x \in \overline{\Omega}} u(x) \geq 0, \\ u(x) < u(x_0'), \quad x \in \Omega. \end{cases}$$

Then the exterior normal derivative $\frac{\partial u}{\partial \mathbf{n}}(x_0')$ of u at x_0', if it exists, satisfies

$$\frac{\partial u}{\partial \mathbf{n}}(x_0') > 0.$$

For a proof of Theorems A.1 and A.2 and a general study of maximum principles, the reader might refer to [PW, Chapter 2] and [Ta2, Chapter 7].

REFERENCES

[Ad] Adams, R.A., *Sobolev spaces*, Academic Press, New York, 1975.

[Ag] Agmon, S., *Lectures on elliptic boundary value problems*, Van Nostrand, Princeton, NJ, 1965.

[ADN] Agmon, S., A. Douglis and L. Nirenberg, Estimates near the boundary for solutions of elliptic partial differential equations satisfying general boundary conditions I, Comm. Pure Appl. Math. **12** (1959), 623–727.

[BL] Bergh, J. and J. Löfström, *Interpolation spaces, an introduction*, Springer-Verlag, Berlin, 1976.

[Bo] Bourdaud, G., L^p-estimates for certain non-regular pseudo-differential operators, Comm. Part. Diff. Eq. **7** (1982), 1023–1033.

[CP] Chazarain, J. et A. Piriou, *Introduction à la théorie des équations aux dérivées partielles linéaires*, Gauthier-Villars, Paris, 1981.

[Fr] Friedman, A., *Partial differential equations*, Holt, Rinehart and Winston, New York, 1969.

[FK] Fujita, H. and T. Kato, On the Navier-Stokes initial value problem I, Arch. Rat. Mech. and Anal. **16** (1964), 269–315.

[Ga] Gagliardo, E., Proprietà di alcune classi di funzioni in più variabili, Ric. di Mat. **7** (1958), 102–137.

[He] Henry, D., *Geometric theory of semilinear parabolic equations*, Lecture Notes in Math. No. 840, Springer-Verlag, Berlin, 1981.

[Ho] Hörmander, L., *The analysis of linear partial differential operators III*, Springer-Verlag, Berlin, Heidelberg, New York, Tokyo, 1985.

[Ku] Kumano-go, H., *Pseudodifferential operators*, MIT Press, Cambridge, Mass., 1981.

[LM] Lions, J.-L. et E. Magenes, *Problèmes aux limites non-homogènes et applications, Vol. 1, 2*, Dunod, Paris, 1968; *Non-homogeneous boundary value problems and applications, Vol. 1, 2*, Springer-Verlag, Berlin, 1972.

[Mi] Mizohata, S., *The theory of partial differential equations*, Cambridge Univ. Press, London, New York, 1973.

[Pa] Pazy, A., *Semigroups of linear operators and applications to partial differential equations*, Springer-Verlag, Berlin, 1983.

[PW] Protter, M. H. and H. F. Weinberger, *Maximum principles in differential equations*, Prentice-Hall, Englewood Cliffs, NJ, 1967.

Typeset by $\mathcal{A}\mathcal{M}\mathcal{S}$-TEX

[RS] Rempel, S. and B.-W. Schulze, *Index theory of elliptic boundary problems*, Akademie-Verlag, Berlin, 1982.

[Se1] Seeley, R.T., Extension of C^∞ functions defined in a half-space, Proc. Amer. Math. Soc. **15** (1964), 625–626.

[Se2] Seeley, R.T., Refinement of the functional calculus of Calderón and Zygmund, Proc. Nederl. Akad. Wetensch., Ser. A **68** (1965), 521–531.

[Se3] Seeley, R.T., Singular integrals and boundary value problems, Amer. J. Math. **88** (1966), 781–809.

[St] Stein, E.M., The characterization of functions arising as potentials II, Bull. Amer. Math. Soc. **68** (1962), 577–582.

[Ta1] Taira, K., On some degenerate oblique derivative problems, J. Fac. Sci. Univ. Tokyo Sec. IA **23** (1976), 259–287.

[Ta2] Taira, K., *Diffusion processes and partial differential equations*, Academic Press, San Diego, New York, London, Tokyo, 1988.

[Ty] Taylor, M., *Pseudodifferential operators*, Princeton Univ. Press, Princeton, NJ, 1981.

[Tr] Triebel, H., *Interpolation theory, function spaces, differential operators*, North-Holland, Amsterdam, 1978.

[Um] Umezu, K., L^p-approach to mixed boundary value problems for second-order elliptic operators, Tokyo J. Math. **17** (1994), 101–123.

[Yo] Yosida, K, *Functional analysis*, Springer-Verlag, Berlin, 1965.

INDEX

Typeset by $\mathcal{A}\mathcal{M}\mathcal{S}$-TEX